EAI/Springer Innovations in Communication and Computing

Series editor
Imrich Chlamtac, European Alliance for Innovation, Gent, Belgium

Editor's Note

The impact of information technologies is creating a new world yet not fully understood. The extent and speed of economic, life style and social changes already perceived in everyday life is hard to estimate without understanding the technological driving forces behind it. This series presents contributed volumes featuring the latest research and development in the various information engineering technologies that play a key role in this process.

The range of topics, focusing primarily on communications and computing engineering include, but are not limited to, wireless networks; mobile communication; design and learning; gaming; interaction; e-health and pervasive healthcare; energy management; smart grids; internet of things; cognitive radio networks; computation; cloud computing; ubiquitous connectivity, and in mode general smart living, smart cities, Internet of Things and more. The series publishes a combination of expanded papers selected from hosted and sponsored European Alliance for Innovation (EAI) conferences that present cutting edge, global research as well as provide new perspectives on traditional related engineering fields. This content, complemented with open calls for contribution of book titles and individual chapters, together maintain Springer's and EAI's high standards of academic excellence. The audience for the books consists of researchers, industry professionals, advanced level students as well as practitioners in related fields of activity include information and communication specialists, security experts, economists, urban planners, doctors, and in general representatives in all those walks of life affected ad contributing to the information revolution.

About EAI

EAI is a grassroots member organization initiated through cooperation between businesses, public, private and government organizations to address the global challenges of Europe's future competitiveness and link the European Research community with its counterparts around the globe. EAI reaches out to hundreds of thousands of individual subscribers on all continents and collaborates with an institutional member base including Fortune 500 companies, government organizations, and educational institutions, provide a free research and innovation platform.

Through its open free membership model EAI promotes a new research and innovation culture based on collaboration, connectivity and recognition of excellence by community.

More information about this series at http://www.springer.com/series/15427

Fadi Al-Turjman
Editor

Trends in Cloud-based IoT

Editor
Fadi Al-Turjman
Research Center for AI and IoT
Near East University
Nicosia, Mersin 10, Turkey

ISSN 2522-8595   ISSN 2522-8609   (electronic)
EAI/Springer Innovations in Communication and Computing
ISBN 978-3-030-40039-2   ISBN 978-3-030-40037-8   (eBook)
https://doi.org/10.1007/978-3-030-40037-8

This Springer imprint is published by the registered company Springer Nature Switzerland AG.
The registered company address is: Gewerbestrasse 11, 6330 Cham, Switzerland

No rain without a cloud.
To my wonderful family and hard working students.
   Fadi Al-Turjman

Preface

We are living in an era where Cloud computing is becoming a global platform for computation and the interaction between humans and machines while performing several critical tasks.

Artificial Intelligence and networking have been considered as a complementary package towards realizing the emerging smart cities paradigm. From this perspective, it is essential to understand the role of these three significant components that will provide a comprehensive vision for the worldwide smart city project in the near future.

No doubt that introducing such a new paradigm can bring significant challenges, especially in terms of overall system performance, cognition and security. It is also essential to consider the emerging intelligent applications for better lifestyle and more optimized solutions in our daily life.

The objective of this book is to present an overview of existing smart city applications while focusing on the issues/challenges surrounding the use of the Cloud. The main focus is on the intelligence aspects that can help in realizing such paradigm in an efficient and secure way. Artificial Intelligent (AI) techniques and new emerging technologies such as the Internet of Things (IoT) accompanied by critical evaluation metrics, constraints and open research issues are included for discussion. This conceptual book, which is unique in the field, will assist researchers and professionals working in the area to better assess the proposed smart cities paradigms, which have already started to appear in our societies.

Hope you enjoy it ...

Fadi Al-Turjman

Contents

About the Editor

 **Fadi Al-Turjman** received his Ph.D. in computer science from Queen's University, Kingston, Ontario, Canada, in 2011. He is a full professor and a research center director at Near East University, Nicosia, Cyprus. Prof. Al-Turjman is a leading authority in the areas of smart/intelligent, wireless, and mobile networks' architectures, protocols, deployments, and performance evaluation. His publication history spans over 250 publications in journals, conferences, patents, books, and book chapters, in addition to numerous keynotes and plenary talks at flagship venues. He has authored and edited more than 25 books about cognition, security, and wireless sensor networks' deployments in smart environments, published by Taylor and Francis, Elsevier, and Springer. He has received several recognitions and best papers' awards at top international conferences. He also received the prestigious *Best Research Paper Award* from Elsevier Computer Communications Journal for the period 2015–2018, in addition to the *Top Researcher Award* for 2018 at Antalya Bilim University, Turkey. Prof. Al-Turjman has led a number of international symposia and workshops in flagship communication society conferences. Currently, he serves as an associate editor and the lead guest/associate editor for several well-reputed journals, including the *IEEE Communications Surveys and Tutorials* (**IF 22.9**) and the Elsevier Sustainable Cities and Society (**IF 4.7**).

Chapter 1
A Blockchain Model for Trustworthiness in the Internet of Things (IoT)-Based Smart-Cities

Rashid Ali, Yazdan Ahmad Qadri, Yousaf Bin Zikria, Fadi Al-Turjman, Byung-Seo Kim, and Sung Won Kim

Our traditional Internet is extended into the Internet of Things (IoT) by connecting physical world devices and objects (things) to it. The diversity of applications of IoT range from mission-critical applications, such as smart-grid, smart-transportation system, smart-surveillance, and smart-healthcare to business applications, such as smart-industries, smart-logistics, smart-banking, smart-insurance, etc. In the IoT revelation, with the assistance of sensors and actuators, conventional devices become smart and autonomous. In coming years, for the predicted huge evolution of the IoT, it is important to provide confidence in this enormous collection of information. Today's advances in the technology are turning IoT revelation into a reality. However, there are still challenges to address, especially for data reliability and security. There is a necessity for inclusive support of privacy and security in IoT for its mission-critical as well as business applications. Numerous frameworks are proposed to enable trustworthiness in the IoT. Academic and industrial researchers are continuously working to propose frameworks for security and privacy in IoT. At the same time, another technology, blockchain has emerged as a key technology that transforms the way in which we share information. Since the time it emerged, this technology has shown auspicious application scenarios. Blockchain builds trust in a

R. Ali (✉)
School of Intelligent Mechatronics Engineering, Sejong University, Seoul, South Korea
e-mail: rashidali@sejong.ac.kr

Y. A. Qadri · Y. B. Zikria · S. W. Kim
Department of Information and Communication Engineering, Yeungnam University, Gyeongsan, South Korea

F. Al-Turjman
Research Center for AI and IoT, Near East University, Nicosia, Mersin 10, Turkey

B.-S. Kim
Department of Computer and Information Communication Engineering, Hongik University, Seoul, South Korea

© Springer Nature Switzerland AG 2020
F. Al-Turjman (ed.), *Trends in Cloud-based IoT*, EAI/Springer Innovations
in Communication and Computing, https://doi.org/10.1007/978-3-030-40037-8_1

distributed atmosphere without the need of central authorities. From the preliminary cryptography to the current smart contracts, blockchain has been functional in many fields. Blockchain as a technological advance has the extreme potential to change many industries, and IoT can be one of them. A blockchain stores all the processed information (known as a transaction in the blockchain terminologies) in sequential order, in a set of memory blocks that are tamper-free to adversaries. All nodes share these processed transactions. The information to these transactions is published as a public ledger that cannot be modified, and every node in the network retains the same ledger. In this way, a blockchain creates a distributed trustworthiness in the network. There have been many studies on the privacy and security issues of blockchain and IoT fusion, however, there lacks a systematic framework on the fusion of blockchain and IoT for trustworthiness. In this chapter, we overview blockchain as a model for IoT trustworthiness. We propose a novel blockchain-based framework for trustworthiness in IoT-based smart-cities.

1.1   Introduction

Internet of Things (IoT) has been extensively grown due to rapid technological advancements to form smart applications such as smart-cities, smart-industries, smart-healthcare and smart-grid, etc. More than 30 billion devices are expected to be part of IoT by 2020 [1]. Connecting billions of objects and things equipped with sensors and actuators will not only improve the quality of experience (QoE) of people but also contribute to the world economy. It is predicted that IoT will contribute more than seven trillion USD to the world economy by 2020 [1]. Although IoT enhances the QoE of users, at the same time, it is vulnerable to a vast number of privacy and security challenges due to massively deployed devices using public communication technologies (such as Wi-Fi, ZigBee, Bluetooth, etc.) [2]. A typical architecture of IoT-based smart-cities is presented in Fig. 1.1. The figure clearly shows that the data generated by multiple sensors in a smart-city travels to the processing layer with the help of different transmission technologies. These privacy and security challenges are known to the manufacturers but either neglected or treated as an afterthought issues [3]. It is critical for the future of IoT that its operational framework is invigorated from expensive, trusted, and centralized architecture to a self-regulating and self-managed decentralized framework [3]. Such a decentralized transformation assures reduced cost of network deployment, scalability of network, self-sufficiency of the environment, security and privacy in a trustless framework, privacy at the user level, and most importantly an access control infrastructure against the network attacks.

However, today we are easy to trust information of financial/industrial and government entities among others without ensuring that the information provided by them and by other external entities, such as IoT industries, has not been tampered or falsified by attackers. Moreover, this is what can be happened if trustworthiness frameworks are centralized architectures. Those untrusted attackers can tamper

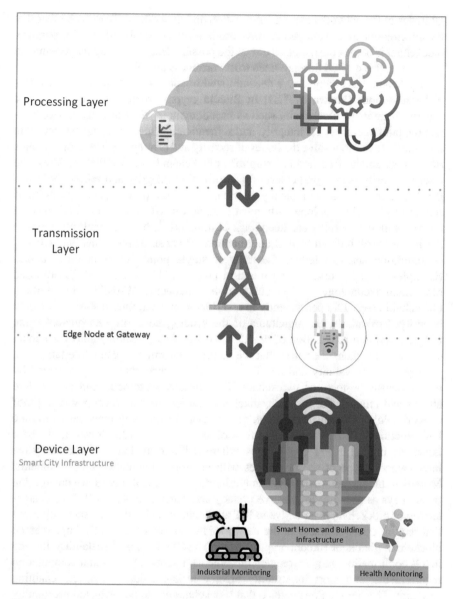

Processing Layer

Transmission
Layer

Edge Node at Gateway

Device Layer
Smart City Infrastructure

Smart Home and Building
Infrastructure

Industrial Monitoring

Health Monitoring

Fig. 1.1 A typical layered architecture of IoT-based smart-cities

information according to their own interests, resulting as a threat to the reliability
of the system. Thus, a framework for trustworthiness is required to verify that the
information has never been modified/tampered. One of the possible solution to bring
trustworthiness in IoT is to implement a distributed model in which all its participant
nodes trust and guarantee that the information remains immutable. For example, if

all nodes in the network have the information and they can verify to each other that the information has not tampered, trustworthiness can be achieved. For this purpose, blockchain is being considered as one of the possible frameworks by the researchers [4–8] to realize desired decentralized trustworthiness in IoT.

Blockchain has emerged as a decentralized trustless security protocol for financial transactions (known as TXs) in Bitcoin cryptocurrency. The cryptographic security benefits of blockchain, such as pseudonymous identities (IDs), decentralization, fault tolerance, TX integrity, and authentication, are attracting researchers to integrate it to IoT to resolve the issues of security and privacy [5]. Therefore, being a distributed, tamper-free, and incorruptible, blockchain has capacities to address the critical security issues, particularly for information integrity and reliability in IoT [9]. IoT architectures are typically distributed, and the peer-to-peer decentralized setting of blockchain is inherently most suitable for IoT architectures. Blockchain record-keeping capability can keep track of transactions between multiple nodes in the IoT network without central coordination. This can deliver a flexible network configuration and can reduce the risks of single point failure in the network. Examples of such works can be found in [10] and [11], where authors proposed blockchain technology for vehicular ad hoc networks (VANET). In addition, blockchain keeps TXs permanently in a verifiable manner, thus ensuring proficient integrity. Particularly, the signatures of the sending nodes in TXs guarantee the integrity and non-repudiation of the TXs. This hashed (pseudonymous IDs) chain structure of blockchain ensures that the recorded information cannot be tampered, even partly. The validity and consistency of the hashed chain are guaranteed by the consensus protocol of blockchain. The consensus protocols can also endure attacks and failures. For instance, attackers with less than 12 hash power in proof of work (PoW) consensus protocol, or less than 13 nodes in practical Byzantine fault tolerance (PBFT) consensus protocol, are tolerated [12]. However, all these issues are critical to IoT applications, where IoT information is generated by the heterogeneous type of sensor devices with heterogeneous network environments. Moreover, blockchain maintains anonymity due to its capability to use changeable public keys as user IDs to preserve privacy and trustworthiness [13]. This characteristic of blockchain is attractive to IoT applications and services, especially those that require trustworthiness in the network [14]. However, currently implemented blockchain (that is of Bitcoin cryptocurrency) has limitations of scalability, latency in TX confirmation, larger storage requirements, intensive power, and computation requirements, and most importantly privacy leakages due to trustless chaining structure. These limitations deduce that blockchain has to be evaluated profoundly before it can be utilized securely and efficiently for trustworthiness in the IoT environment.

Recently, researchers have published a numerous research work on blockchain-based IoT technology [4–9]. These works either focus on general applications of blockchain in IoT or discuss technical aspects concerning only about the digital cryptocurrencies. Their work does not give an insight into blockchain challenges related to trustworthiness in IoT. For example, Yli-Huumo et al. in [15], discussed the issues of distributed denial of service (DDoS) attacks, 51% attack, information malleability, authentication and usability problems from cryptocurrencies (Bitcoin) point of view. Similarly, in [16], authors carry out a detailed survey of blockchain technologies and their influence on advancements. It also highlights the issues associated with Bitcoin, and draw attention towards the prospective utilization of blockchain. However, this work discusses IoT as a short subsection in the long list of potential use cases of the blockchain. In [17], authors discuss multiple variants of blockchain, such as Ethereum [18], Ripple [19], Gridcoin [20] etc., and wrote an essence of few of the applications of the blockchain. Their work also lacks the discussion of issues concerning the integration of blockchain and IoT. Authors in [21] present a lightweight blockchain framework for smart-home IoT application. Their proposed work proposes a solution to avoid Bitcoin's issues of intensive computation, the latency of TX confirmation, and blockchain scalability. Huh et al. [22] proposed one of the use cases for blockchain and IoT integration, which configure and manage IoT devices with the help of blockchain smart contracts. Their proposed architecture aims to resolve the security and synchronization issues in a client–server IoT network. In their proposed solution, all the connected IoT devices become vulnerable to security threat if the server gets malicious. Ethereum smart contracts were suggested by the authors of [18] to take advantage of blockchain trust-free decentralized framework for the IoT devices. In addition, Conoscenti et al. [23] carried out a detailed literature review of blockchain applications beyond cryptocurrencies and presented their fusion to IoT-based applications.

Therefore, to enable the blockchain adoption in IoT networks for trustworthiness, there is a requirement of carrying out a deeply dived overview of these two technologies to find out how does existing blockchain technology can help for IoT trustworthiness. Moreover, the challenges for IoT are to leverage blockchain to resolve its security and privacy issues, and to find out the impediments in doing so. Thus, in this chapter, we carried out a precise review of blockchain technology for threat environment and resultant IoT security and privacy requirement. We also discuss two of the blockchain technologies: smart contracts and consensuses mechanisms to determine a suitable framework for IoT trustworthiness. It is presumed that the implementation of these blockchain technologies at the cloud networking and communication layers meets most of the IoT trustworthiness requirements such as device authentication and authorization, device identity management, information confidentiality, low latency in frame transmission confirmation (acknowledgment).

The rest of the chapter is organized as follows: Sect. 1.2 provides a deeply dived overview of blockchain for IoT. In Sect. 1.3, major blockchain technologies, smart contract and consensus mechanism, are discussed. Section 1.4 presents our proposed blockchain model for IoT trustworthiness in smart-cities. Finally, the chapter is concluded with a hint of future work in Sect. 1.5.

1.2   Blockchain for IoT: An Overview

In this section, we first present an introduction of blockchain technology, and then explain how it works? Later subsections of this section include types of blockchain, terminologies in blockchain, feature and applications of blockchain, and finally the challenges and limitations of the blockchain technology. The purpose of this section is to briefly introduce blockchain technology before discussing its fusion to the IoT for trustworthiness development.

1.2.1   What Is Blockchain?

Trust management in distributed information systems is tremendously complex due to the absence of verification/audit mechanisms, especially in case of sensitive information, such as financial transactions with digital currencies. In 2008, Satoshi Nakamoto [24] introduced Bitcoin as a virtual cryptocurrency, which works without the aid of any centralized authority that works in a decentralized manner. Bitcoin is detained securely in a decentralized peer-to-peer (P2P) auditable and verifiable network of nodes (known as actors). Besides the Bitcoin, Satoshi Nakamoto presented the idea of blockchain, whose acceptance has arisen even further than the cryptocurrencies. As discussed earlier, blockchain is a protocol to verify the transactions by a group of untrustworthy users (actors). It is a distributed, immutable, transparent, secure, and auditable ledger. The real magnificence of blockchain is that it can be consulted freely and fully, and allows access to all the previous transactions until the first transaction of the system. All the transactions can be checked and verified by any actor at any time.

1.2.2   How Blockchain Works?

A very fundamental structure of blockchain can be divided into three layers, that is, underlying P2P network of nodes, global ledger, and various applications such as Bitcoin and Ethereum [25] as shown in Fig. 1.2. The lowest layer that is of P2P network nodes is accountable to guarantee spontaneous communication among the nodes/actors where these actors are physically distributed. Though these actors are geographically dispersed, they give equal privileges to participants in the application. As we can see from the figure, the P2P network layer is decentralized, where each node is an information provider as well as an informed consumer. These nodes are responsible for the routing process in the network, in which they discover and maintain the connections between the neighboring peer nodes, and hence propagate and verify the transactions in the network. In other words, nodes in the network synchronize the information blocks (data structure) of the blockchain

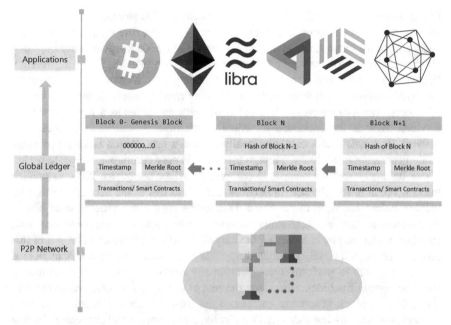

Fig. 1.2 Fundamental layered structure of the blockchain

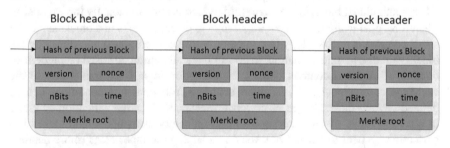

Fig. 1.3 The data structure of the blockchain

and are the key reflection and basis of the blockchain decentralization characteristic. The data structure of blockchain contains information of TXs performed at a given time in a chain of blocks. Each block in the chain is linked together with its previous block by a reference ID (previous block header hash) as shown in Fig. 1.3. Peers in the network provide functionalities of routing, storage, wallets services, and mining of the TXs to support and operate with the blockchain [26]. Based on the functionalities, nodes are divided into four categories: Bitcoin core nodes, full nodes, solo minor nodes, and light wallet nodes. Bitcoin core nodes have the functionalities of routing, storage, wallet, and mining. Full nodes are capable of routing and storage only. Similarly, solo minor nodes have functionalities of routing, storage, and mining. And lastly, the light wallets can perform routing and wallet services only. The transaction and block propagation are included in routing function, thus it is very necessary for a P2P network. The storage function keeps a copy of the chain in the node, however, the copy of the entire chain for full nodes and a part of the chain for light nodes. The wallet in a node keeps security keys that allow users to perform transactions in the network (such as to handle their Bitcoins). Finally, the mining function takes care of adding new blocks by implementing the PoW consensus. All these nodes performing mining functionalities are known as miners. Miners receive newly created Bitcoins and a fee as a reward for their services. PoW is the backbone for enabling trustless consensus in a blockchain network. It consists of a computationally intensive task, which is very necessary while generating new blocks. Though this work is very complex to solve, however, it can be easily verified once it is completed. Miner completes PoW to publish the new block in the blockchain network, afterward rest of the network verifies its validity before adding it to the chain. Due to the concurrent generation of blocks by different miners, blockchain may temporarily fork into different branches. This situation is resolved by considering the longest branch of blocks as valid. All of this process results in a novel, distributed trustless consensus mechanism. A malicious attacker requires to perform a computationally expensive procedure to tamper a block to corrupt the blockchain since the rest of the trusted miners would outstrip the malicious block generation and hence the trusted branch of blocks will undo the one generated by the attacker. Technically, a malicious attacker requires having control of at least 51% of the computing resources of the network if he/she wants to add a manipulated block to the chain [26], which is computationally too expensive. Therefore, practically it is almost impossible to corrupt blocks in a blockchain. The smart contract is a technology (a computer protocol or program) introduced by the blockchain that permits a contract to be automatically enforced based on predefined conditions. A smart contract defines the execution logic for a transaction in the exchange of cryptocurrency. In smart contracts, functions and conditions can be defined beyond the exchange of cryptocurrencies, such as the validation of assets in a certain range of transactions with non-monetary elements, which makes it a perfect component to expand blockchain technology to other areas (more detailed description of smart contracts is in later sections). One of the pioneer blockchains that included smart contracts is Ethereum [25], and today smart contracts have been introduced in many of existing blockchain implementations.

Types of Blockchain

Blockchains are categorized into either permissioned or permissionless categories. Based on these categories, there are multiple types of blockchains, such as public blockchain, private blockchain, hybrid blockchain, and consortium blockchain. The key differences between different types of blockchains are shown in Table 1.1. A blockchain can be either permissioned or permissionless based on the restrictions to creating new blocks of TXs. If a blockchain is permissionless, any node can create new blocks of TXs, however, in a permissioned blockchain, only selected nodes can

Table 1.1 Comparison between permissioned and permissionless blockchains

Permissioned blockchain	Permissionless blockchain
Requires permission before participation	Does not require permission before participation
Participants are already known	Participants are unknown
Limited number of participants	Unlimited number of participants
Data security	Less data privacy
Instant consensus inevitability	Week consensus inevitability
High transaction throughput	Low transaction throughput
Less scalable	High scalable
Vulnerability for participant collusion	51% attack vulnerable

process TXs. On the other hand, types of blockchain only concern about access to the blockchain data [27].

Public Blockchain

Usually, public blockchain is a permissionless blockchain that allows free and unconditional access to the digital ledger by any node in the network [27]. To encourage the miners for adding new blocks to the public blockchain (mining) is mostly incentive-based. The connectivity between the nodes in public blockchain takes longer time to finalize the TXs. The transparency in the network is maintained by making all the TXs visible to the public (nodes in the network). It is obvious that the issues and challenges of user data privacy emerge. Since there exists poor TX finality, public blockchain has low TX throughput (usually measured in transactions per seconds TPS), especially when PoW is employed [28]. Bitcoin, Ethereum, Litecoin [29], and Lisk [30] are some of the examples of public blockchains.

Private Blockchain

A private blockchain is usually a permissioned ledger of a limited number of miners with known IDs. Therefore, this type of blockchain has restricted TX processing to the predefined miners. Moreover, a miner can only access those TXs that are directly related to it [27]. Hyperledger-Fabric [31] is one of the examples of private blockchain, in which miners maintain the privacy of their TXs using private channels. These private channels are known as restricted communication paths that provide TX privacy and confidentiality for the specific number of miners. Thus, private blockchains are more trustful as compared to public blockchains. Since the number of authorized miners is very less in case of a private blockchain, these are faster than the public blockchains and has higher throughput as well as [28]. However, those private blockchains using Byzantine fault tolerance (BFT) consensus suffer poor scalability issues due to less number of validators. Zheng et al.

[9] added that the TX blocks in private blockchain can be tampered due to its partial centralization. Some of the examples of private blockchains include: Hyperledger, Multichain [32], Quorum [33], etc.

Hybrid/Consortium Blockchain

Hybrid blockchain is also known as partially decentralized or consortium blockchain and is a balance between private and public blockchain [17]. In this type of blockchain, few of the pre-selected nodes control the consensus process. For example, a Hyperledger, wherein a consortium of few industries maintains a mining node in the blockchain network. A block is only validated if at least seven nodes have signed it. Other nodes in the network may have open read-only access to the blockchain or can be restricted [18]. Therefore, a consortium blockchain is also called partially decentralized. However, due to abridged decentralization, consortium blockchain can easily tamper [9].

1.3   Major Blockchain Technologies: Smart Contract and Consensus Mechanisms

In this section, we first present a recently emerged technology of blockchain, a smart contract that can perform arbitrary computations other than transferring cryptocurrencies. After that, we present the fundamental trust mechanism used in a blockchain, that is, consensus mechanism.

1.3.1   Smart Contract

It has been already a lot we mentioned this term in the paper, and now is the time to dig it into detail. A smart contract is a protocol containing terms and conditions for the computerized transaction of a contract. It has a way to impose or execute contractual clauses. The concept of smart contract was technologically unviable in the beginning. However, blockchain technology turned out to be the best technology for smart contract protocol. The integration of smart contract into blockchain has significant contributions and has headed to the next generation of blockchain technology, Blockchain 2.0. These automatically performed contracts in a decentralized environment promise to change the way current transactions are working.

The code of smart contract resides in the blockchain and identified by a unique ID (address). The users send their transactions using this address of a smart contract. The contract is executed by the consensus mechanism of the blockchain (for further details of consensus, see next section). The implementation of smart contracts in blockchain brings advantages such as reduced cost, faster transaction processing, precision, efficiency, and transparency. It has fostered the appearance of many of the new applications in a wide variety of areas. Bitcoin uses a very basic scripting language, thus it is inadequate to support the technology like a smart contract. This has led to the emergence of new blockchain platforms in which smart contract functionality is already integrated, such as Ethereum. Ethereum is one of the most prominent smart contract blockchain platforms. It has a built-in Turing-complete programming language that defines the smart contracts and decentralized applications. Smart contracts access data about different states and events provided by the oracles. These objects are vital for the effective integration of smart contracts within the real world. However, they bring extra complexity because of trust provisioning in the oracles [34]. Therefore, smart contracts are vulnerable to many of the network attacks, and brings novel exciting challenges and issue to consider [35]. Another problem with smart contracts is that if bugs are found in the contract code, it is a very critical issue because it is irreparable and unchallengeable. Moreover, real-world contracts usually have terms and conditions that are not assessable. Therefore, there is still a lot of work to be done in order to model the conditions of the contracts in smart contracts, so that they are representable and quantifiable for a network device to execute them [36].

1.3.2   Consensus Mechanism

The consensus mechanism is one of the vital parts of the blockchain network and is responsible for the trustworthiness of the information contained in blockchain [37]. The main objective of the consensus mechanism is to implement agreement rules in a decentralized trustless network. There are many types of consensus mechanisms used by different blockchain networks. Such as Proof of Work (PoW), Proof of Service (PoS), and Practical Byzantine Fault Tolerance (PBFT), etc. Some of the prominent consensus mechanisms are discussed in the following paras.

Proof of Work (PoW)

PoW was the very first consensus introduced by the Bitcoin, which forces miner to solve a computationally very intensive but easy to verify task before creating a new block. The solution to the task is published after it is solved and the new block is added to the corresponding chain. Later, this newly added block is spread across the network for verification from the rest of the nodes and append it to its blockchain [38]. The process of adding new blocks concurrently happens in multiple parts of the network, thus forming a hierarchical tree of the chain. In the result, there exist several valid branches simultaneously in the blockchain. Before adding a new block, each peer in the network verifies that the branch they are adding to is the longest chain, which is assumed valid. The key issue with this type of consensus is that it consumes high energy due to extensive computations. As mentioned earlier, the 51% attack is the possible attack on the Bitcoin. Moreover, inducements in PoW are surprisingly endorsing centralization as the spread of mining pools confirms.

Several drawbacks including high latency, low transaction rate, and high-energy consumption make PoW unsuitable for numerous of the applications, especially energy constraint applications. There have been many attempts by the researchers to modify PoW. Some of them are as follows.

Proof of Service (PoS)

One of the popular alternative approaches to PoW consensus in a blockchain network is the PoS. PoS is saved on the fact that the user with more coins is more interested in the survival of the chain and the correct functioning of the network, and therefore are a more appropriate response for the protection of the network. The main objective of the use of PoS is to change the prospective expense from outside the network to inside the network. One of the problems with this consensus, also known as nothing-at-stake, is that this approach does not provide any incentives to the minors for their block creation. Hence, it promotes the enrichment of the existing rich. The examples of the blockchain using this type of consensus are PeerCoin [39] and Ethereum. BitShare [40], one of the blockchain, used a variant of PoS known as delegate PoS (DPoS). In DPoS, a number of nominated witnesses authorize signatures and time stamps of transactions and include them in the blockchain network. The witness/user is rewarded each time a new block is successfully added to the blockchain. Faster transaction confirmation is achieved by allowing delegates to set the block latency and size.

Leased Proof of Stake (LPoS)

In LPoS [41], a user is allowed to lease funds to other users so that they can be trusted and selected for new block creation. It increases the number of electable users in the network, and hence reduces the probability of the blockchain network being controlled by a single group of participants. In this type of consensus mechanism, all the participants share their rewards proportionally. One of the blockchains known as Waves [41] uses LPoS.

Proof of Burn (PoB)

Another consensus mechanism, known as PoB [42] proposes burning coins by sending them to a verifiable non-spendable address in order to publish a new block. PoB is very similar to PoW and is hard to compute while easy to verify. However, it is very energy consumption [42]. Additionally, it provides financial suggestions that contribute to a steadier environment. In this protocol, users burn their coins in order to mine blocks into the blockchain. PoB reforms the expense that users have to tolerate in order to mine in the blockchain. As users who burn their coins into the blockchain will be devoted in order to have more rewards from mining, PoB encourages long-term involvement of the user in a project. Thus, a high percentage of long-term users in a blockchain might influence the price stability of these coins. One of the challenges to PoB is that there is a lot of wastage of resources and most of the mining power belongs to those miners who burn more coins. Slimcoin [43] uses PoB as the consensus of its blockchain protocol.

Proof of Importance (PoI)

PoI [44] associates an importance value with each user to build a reputation system in the network. The chance of the user being chosen to create a block depends upon this value. The computation of this value also takes into account the number of used coins and the number of successful transactions. This consensus mechanism determines the users that are qualified to add a block to the blockchain through a process called harvesting. The users that harvest a block into the blockchain collect transaction fees associated with the block. Higher the importance value is higher will be the chance for a user to be chosen for harvesting the blocks. PoI is a user's support-based network, where users are encouraged to spend their coins and spread them into the network. It is a contrary algorithm to the PoS, which is said to support hoarders. This consensus protocol was introduced by NEM [44] blockchain.

Proof of Activity (PoA)

A new hybrid of PoW and PoS protocol for cryptocurrency was proposed in [45], named as Proof of Activity (PoA). PoA suggests good security against possibly practical attacks on a blockchain and has a somewhat small consequence in terms of communication and storage. In PoA, mining of a block begins in the traditional manner used by PoW, that is, miners vying to be the first to solve a puzzle and claim their rewards [45]. The only difference is that the successfully mined blocks do not enclose TXs. Instead, they are only templates including header information and the reward address. After the successful mining of this nearly blank block, PoA switches utilize PoS protocol, where the header information is used to select a random group of validators to sign the block for the blockchain. The nodes discard a block as incomplete if it persists unsigned by some or all of the selected validators after a given time. However, PoA has been criticized, as it still requires a large number of resources for mining a single block. The double signing of the block by a single miner also remains an issue with PoA [45]. Espers [46] is the cryptocurrency that uses PoA consensus mechanism.

Proof of Elapsed Time (PoET)

A blockchain platform developed by Intel known as Intel Sawtooth Lake (also known as IntelLedger) introduced PoET consensus mechanism [38]. PoET was intended to execute in a trusted execution environment (TEE), such as Intel's Software Guard Extension (SGX) [47]. The PoET uses a random leader election model (also referred to as a lottery-based election model), where the next leader is randomly selected by the protocol to finalize the block. Later, it deals with the untrusted and open-ended users in the consensus algorithm. In this consensus mechanism, all miners have to run the TEE using Intel SGX to request a wait time from the code running inside the TEE. The miner with the shortest waiting time becomes the leader. A validated miner that claims to be the leader for block creation can also produce proof generated with the TEE, which is verifiable, by all other users. The drawback of this consensus mechanism is that it relies on specialized hardware.

Practical Byzantine Fault Tolerance (PBFT)

Hyperledger-Fabric [31] is one of the popular blockchains that offers a flexible framework with a pluggable consensus mechanism. Hyperledger-Fabric supports PBFT, which executes non-deterministic blockchain codes. PBFT was proposed as a first practical solution to the Byzantine failures [48, 49]. The concept of a replicated state machine and voting by replicas are used by PBFT for blocks validation. It has many features of optimization, such as validation and encryption of TXs between replicas and users, and reduces the size and number of TXs, for the system to practically face the Byzantine faults. According to PBFT consensus mechanism, 3f+1 replicas are required to tolerate "f" failing nodes. In this mechanism, the overhead on the performance of the replicated services is reduced. However, its TXs overhead increases significantly as the number of replicas increases [50].

Consequently, there has been considerable work done to propose consensus mechanisms for different blockchain platforms. However, there is still a lack of work done that proposes these two technologies to be used for IoT trustworthiness.

1.4   Proposed Blockchain Model for IoT Trustworthiness

In this section, we propose a blockchain-based model for trustworthiness in IoT-based smart-cities. Our proposed model suggests two scenarios for using blockchain technology to provide IoT network trustworthiness (as shown in Fig. 1.4). In the first scenario, an IoT service provider integrates IoT devices (sensors, wearable, etc.) to receive and transmit data and connects these devices to a blockchain network. At this point, blockchain technology provides the ability for the connected IoT devices to exchange messages, make orders, and complete transactions (TXs). The trust-

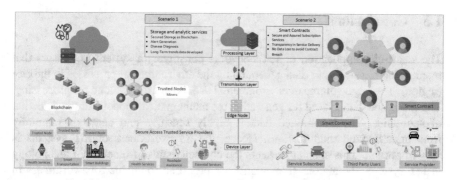

Fig. 1.4 Proposed blockchain model for trustworthiness in IoT-based smart-cities

worthiness of the blockchain will mitigate the security attacks because the attacker uses trustless backdoors to attack the network [51]. The client nodes (sensors) are connected to the gateway which has sufficient energy and memory resources. Therefore, the gateway acts as an edge node. The blockchain is implemented at the edge node, which keeps the record of all the sensor nodes that join the network and who are authorized to send and receive data. If a malicious node tries to impersonate an authentic node, the blockchain is able to identify the malicious node. Additionally, all the sensor nodes keep track of the data traffic sent by them and update the gateway periodically. Since the gateway also knows the amount of data it receives from each client, it is able to detect any traffic pattern anomalies and also localize the source of the problem. The cloud also maintains a larger blockchain that implements the same principles for enhancing the scalability of the system. Additionally, the stored data is processed using Big Data Analytics to obtain a diagnosis for the user. Additionally, the analysis of the traffic flow from the gateway blockchain can help in real-time detection of security attacks.

In the second proposed scenario, Ethereum smart contracts can be implemented into the IoT system in order to automate and regulate the smart-city services subscribed by the trustful nodes. As described earlier, Ethereum is a global, open-source blockchain platform for decentralized IoT applications, which provides the facility to write code that controls the digital information, runs exactly as it is programmed, and can be accessed from everywhere over the Internet. The reason we propose the use of Ethereum into the IoT systems for trustworthiness issues is that it will provide a seamless and safe exchange of TXs between connected nodes just as it performs in blockchain-based digital cryptocurrencies [8]. The services that are subscribed by the user are regulated by the terms and conditions of the smart contract and if there are any violations of the terms and conditions, remedial actions can be automatically put into motion.

1.5 Conclusions

The smart-city application of IoT networks constitutes a system of sensors that monitor the physio-cyber data generated by the environmental processes in the city (region). The large volume of environmental data is sent to the cloud for processing. The processed data is used for generating the mathematical models for the detection of anomalies. The transmitted data is highly sensitive and demands immunity from unauthorized modifications. Hence, the trustworthiness of IoT systems is very important. This chapter presents the characteristics of blockchain technology in IoT systems. Numerous frameworks are proposed to enable trustworthiness in the IoT. Academic and industrial researchers are continuously working to propose frameworks for security and privacy in IoT. At the same time, another technology, blockchain has emerged as a key technology that transforms the way in which we share information. Since the time it emerged, this technology has shown auspicious application scenarios. Blockchain builds trust in a distributed atmosphere without the need of central authorities. From the preliminary cryptography to the current smart contracts, blockchain has been functional in many fields. Blockchain as a technological advance has the extreme potential to change many industries, and IoT can be one of them. A blockchain stores all the processed information (known as a transaction in the blockchain terminologies) in sequential order, in a set of memory blocks that are tamper-free to adversaries. All nodes share these processed transactions. The information to these transactions is published as a public ledger that cannot be modified, and every node in the network retains the same ledger. In this way, a blockchain creates a distributed trustworthiness in the network. There have been many studies on the privacy and security issues of blockchain and IoT fusion, however, there lacks a systematic framework on the fusion of blockchain and IoT for trustworthiness. In this chapter, we propose to use decentralized and trustworthiness capabilities of Blockchain as a framework for IoT trustworthiness in smart-cities. The proposed model suggests using two scenarios for the implementation of blockchain for trustworthiness in IoT.

References

1. D. Lund, C. MacGillivray, V. Turner et al., Worldwide and Regional Internet of Things (IoT) 2014-2020 Forecast: a Virtuous Circle of Proven Value and Demand. International Data Corporation (IDC), Technical Report (2014)
2. Y.B. Zikria, S.W. Kim, O. Hahm et al. Internet of Things (IoT) operating systems management: opportunities, challenges, and solution. Sensors 19(8), 1–10 (2019)
3. J. Wurm, K. Hoang, O. Arias et al., Security analysis on consumer and industrial IoT devices, in *The IEEE 21st Asia and South Pacific Design Automation Conference (ASP-DAC)*, Macau, China, pp. 519–524 (2016)

4. D. Minoli, B. Occhiogrosso, Blockchain mechanisms for IoT security. Internet Things. **1**(2), 1–13 (2018)
5. A. Reyna, C. Martín, J. Chen et al., On blockchain and its integration with IoT. Challenges and opportunities. Futur. Gener. Comput. Syst. **88**, 173–190 (2018)
6. I. Makhdoom, M. Abolhasan, H. Abbas et al., Blockchain's adoption in IoT: the challenges, and a way forward. J. Netw. Comput. Appl. **123**, 251–279 (2019)
7. K. Christidis, M. Devetsikiotis, Blockchains and smart contracts for the internet of things. IEEE Access. **4**, 2292–2303 (2016)
8. M.S. Ali, M. Vecchio, M. Pincheira et al., Applications of blockchains in the internet of things: a comprehensive survey. IEEE Commun. Surv. Tutorials. **21**(2), 1676–1717 (2019)
9. Z. Zheng, S. Xie, H.N. Dai et al., Blockchain challenges and opportunities: a survey. Int. J. Web Grid Serv. **14**, 352–375 (2018)
10. P.K. Sharma, S.Y. Moon, J.H. Park, Block-VN: a distributed blockchain based vehicular network architecture in smart city. J. Inf. Process. Syst. **13**(1), 184–195 (2017)
11. B. Leiding, P. Memarmoshrefi, D. Hogrefe, *Self-Managed and Blockchain-Based Vehicular Ad-hoc Networks*. (UbiComp Adjunct, Heidelberg, 2016), pp. 137–140
12. M. Castro, B. Liskov, Practical Byzantine fault tolerance, in *3rd Symposium Operating Systems Design Implementation, OSDI'99*, New Orleans, pp. 173–186 (1999)
13. M.C.K. Khalilov, A. Levi, A survey on anonymity and privacy in Bitcoin-like digital cash systems. IEEE Commun. Surv. Tutorials. **20**(3), 2543–2585 (2018)
14. X. Zha, K. Zheng, D. Zhang, Anti-pollution source location privacy preserving scheme in wireless sensor networks, in *13th Annual IEEE International Conference on Sensing, Communication, and Networking (SECON)*, London, pp. 1–8 (2016)
15. J. Yli-Huumo, D. Ko, S. Choi et al., Where is current research on blockchain technology? A systematic review. PLoS ONE. **11**(10), 0163477 (2016)
16. Institute NR, Survey on Blockchain Technologies and Related Services (2016), Available from: https://www.meti.go.jp/english/press/2016/pdf/0531_01f.pdf. [Cited 2019 May 22]
17. M. Pilkington, Blockchain technology: principles and applications, *Research Handbook on Digital Transformations* (Edward Elgar Publishing, Cheltenham, 2015), p. 35
18. V. Buterin, A Next-Generation Smart Contract and Decentralized Application Platform. Ethereum White Paper (2014), pp. 1–36
19. XRP: The Digital Asset for Payments [homepage on the Internet]. Ripple (2013), Available from: https://ripple.com/xrp/. [Cited 2019 May 23]
20. Gridcoin, The Computational Power of a Blockchain Driving Science & Data Analysis. Gridcoin White Paper (2018), pp. 1–12.
21. A. Dorri, S.S. Kanhere, R. Jurdak, Blockchain in internet of things: challenges and solutions. CoRR. arXiv:1608.05187 (2016). Available from: http://arxiv.org/abs/1608.05187
22. S. Huh, S. Cho, S. Kim, Managing IoT devices using blockchain platform, in *IEEE 19th International Conference on Advanced Communication Technology (ICACT)* (IEEE, Bongpyeong, 2017), pp. 464–467
23. M. Conoscenti, A. Vetrò, J. Carlos De Martin, Blockchain for the internet of things: a systematic literature review, in *IEEE/ACS 13th International Conference of Computer Systems and Applications (AICCSA)* (IEEE, Agadir, 2016), pp. 1–6
24. S. Nakamoto, Bitcoin: A Peer-to-Peer Electronic Cash System. Online Unpublished Paper (2008), pp. 1–9. Available from: https://bitcoin.org/bitcoin.pdf
25. Q. Feng, D. He, S. Zeadally et al., A survey on privacy protection in blockchain system. J. Netw. Comput. Appl. **126**, 45–58 (2019)
26. A.M. Antonopoulos, *Mastering Bitcoin: Unlocking Digital Cryptocurrencies* (O'Reilly Media Inc., Sebastopol CA, 2014)
27. J. Garzik, Public Versus Private Blockchains Part 1. Permissioned Blockchains. Semantic Scholar. (2015). Available from: https://www.semanticscholar.org/paper/Public-versus-Private-Blockchains-Part-1-%3A/f5c596ada0674449879965c5cc4c347c2b2b3180

28. K. Lukas, In-depth on Differences Between Public, Private and Permissioned Blockchains. Online unpublished. (2015). Available from: https://medium.com/@lkolisko/in-depth-on-differences-between-public-private-and-permissioned-blockchains

29. Litecoin: The Cryptocurrency for Payments [homepage on the Internet]. Litecoin (2011). Available from: https://litecoin.org/. [Cited 2019 May 25]

30. Lisk Documentation [homepage on the Internet]. Lisk (2018). Available from: https://lisk.io/documentation. [Cited 2019 May 25]

31. Hyperledger, Hyperledger-Fabricdocs Documentation [homepage on the Internet] (2019). Available from: https://media.readthedocs.org/pdf/hyperledger-fabric/latest/hyperledger-fabric.pdf. [Cited 2019 June 27]

32. G. Gideon, Multichain Private Blockchain. White Paper. (2015), pp. 1–17. Available from: https://www.multichain.com/download/MultiChain-White-Paper.pdf

33. Quorum, Quorum White Paper (2016), pp. 1–8. Available from: https://github.com/jpmorganchase/quorum-docs/blob/master/Quorum%20Whitepaper%20v0.1.pdf

34. F. Zhang, E. Cecchetti, K. Croman et al., Town crier: an authenticated data feed for smart contracts, in *2016 ACM SIGSAC Conference on Computer and Communications Security*, Vienna, pp. 270–282 (2016)

35. K. Delmolino, M. Arnett, A. Kosba et al., *Step by Step Towards Creating a Safe Smart Contract: Lessons and Insights from a Cryptocurrency Lab* (Springer, Berlin/Heidelberg, 2016)

36. C.K. Frantz, M. Nowostawski, From institutions to code: towards automated generation of smart contracts, in *IEEE International Workshops on Foundations and Applications of Self Systems*, Augsburg, pp. 1–6 (2016)

37. C. Cachin, M. Vukolic, Blockchains Consensus Protocols in the Wild. arXiv preprint. (2017), pp. 1–24. Available from: https://arxiv.org/pdf/1707.01873.pdf

38. A. Baliga, Understanding Blockchain Consensus Models. Persistent Systems: Whitepaper (2017), pp. 1–14. Available from: https://www.persistent.com/wp-content/uploads/2017/04/WP-Understanding-Blockchain-Consensus-Models.pdf

39. S. King, S. Nadal, Peercoin-Secure and Sustainable Cryptocoin. Online Whitepaper (2012), pp. 1–15. Available from: https://peercoin.net/whitepaper

40. F. Schuh, D. Larimer, BitShares 2.0: General overview. Cryptonomex: Whitepaper (2017), pp. 1–10. Available from: https://cryptorating.eu/whitepapers/BitShares/bitshares-general.pdf

41. Leased Proof-of-Stake (LPoS) [homepage on the Internet]. Cryptographics; 2017–2019. Available from: https://cryptographics.info/cryptographics/blockchain/consensus-mechanisms/leased-proof-stake/. [Cited 2019 June 28]

42. J. Frankenfield, Proof of Burn (Cryptocurrency) [homepage on the Internet]. Investopedia (2018). Available from: https://www.investopedia.com/terms/p/proof-burn-cryptocurrency.asp. [Cited 2019 June 28]

43. P4Titan, Slimcoin: A Peer-to-Peer Crypto-Currency with Proof-of-Burn "Mining without Powerful Hardware". Whitepaper (2014), pp. 1–6. Available from: http://www.doc.ic.ac.uk/~ids/realdotdot/crypto_papers_etc_worth_reading/proof_of_burn/slimcoin_whitepaper.pdf

44. A. Nember, NEM Technical Reference. Whitepaper (2018), pp. 1–58. Available from: https://nem.io/wp-content/themes/nem/files/NEM_techRef.pdf

45. I. Bentov, C. Lee, A. Mizrahi et al., Proof of activity: extending Bitcoin's proof of work via proof of stake. SIGMETRICS Perform. Eval. Rev. **42**, 34–37 (2014). Available from: http://netecon.seas.harvard.edu/NetEcon14/Papers/Bentov_netecon14.pdf

46. Batysta, Espers: Cryptocurrency with Hybrid PoW/PoS and Unique Algorithm [homepage on the Internet]. Espers (2017). Available from: https://blog.espers.io/espers-cryptocurrency-with-hybrid-pow-pos-and-unique-algorithm-63da942e307d. [Cited 2019 June 28]

47. Intel Corporation, Intel's Software Guard Extensions (SGX): The control of protecting your data [homepage on the Internet] (2019). Available from: https://www.intel.com/content/www/us/en/architecture-and-technology/software-guard-extensions.html. [Cited 2019 June 28]

48. V. Gramoli, From blockchain consensus back to Byzantine consensus. Futur. Gener. Comput. Syst. **1**(1), 1–10 (2017)

49. F. Knirsch, A. Unterweger, G, Eibl et al., *Sustainable Cloud and Energy Services: Principles and Practices* (Springer, Cham, 2017), pp. 85–116
50. J. Zhu, P. Liu, L. He, Mining information on Bitcoin network data, in *IEEE International Conference on Internet of Things (iThings) and IEEE Green Computing and Communications (GreenCom) and IEEE Cyber, Physical and Social Computing (CPSCom) and IEEE Smart Data (SmartData)*, pp. 999–1003 (2017)
51. I. Tomic, J.A. McCann, A survey of potential security issues in existing wireless sensor network protocols. IEEE Internet Things J. **4**(6), 1910–1923 (2017)

Chapter 2
A Review on the Use of Wireless Sensor Networks in Cultural Heritage: Communication Technologies, Requirements, and Challenges

Hadi Zahmatkesh and Fadi Al-Turjman

2.1 Introduction

Smart cities [1] use information and communication technologies (ICTs) related to the Internet of Things (IoT) [2] to monitor and supervise the current resources and infrastructures. Intelligent management of the cultural heritage (CH) in smart cities is a service which protects the CH from degradation [3] since the CHs are valuable and important objects for smart cities from the touristic point of view [4]. Natural disasters, bad environmental conditions, and human interventions are gradually weakening the health of the CH such as historical buildings, museums, and exhibitions. Therefore, it is of the utmost importance to monitor the environment in order to protect the CH from degradation. In recent years, real-time monitoring of the CH and artworks has attracted too much attention by both academia and industry [5, 6]. Continuous monitoring of environmental conditions is necessary for appropriate protection of the CH such as artworks, museums, ancient books, historical buildings, and artifacts. Various environmental parameters such as temperature, humidity, and light have to be monitored and kept within a predefined range recommended by experts [7]. In this regard, wireless sensor network (WSN) plays a significant role in preserving and protecting the CH, and provides continuous and real-time monitoring of the environment. Moreover, WSNs can have an important role in informing the authorities or, if possible, automatically changing the environment's conditions by using heating, ventilation, and air condition (HVAC) systems [8].

H. Zahmatkesh (✉)
Department of Mechanical, Electronic and Chemical Engineering, OsloMet—Oslo Metropolitan University, Oslo, Norway

F. Al-Turjman
Research Center for AI and IoT, Near East University, Nicosia, Mersin 10, Turkey

© Springer Nature Switzerland AG 2020
F. Al-Turjman (ed.), *Trends in Cloud-based IoT*, EAI/Springer Innovations in Communication and Computing, https://doi.org/10.1007/978-3-030-40037-8_2

This chapter provides a comprehensive review of the use of WSNs in the CH context. It starts with a highlight of the current literature in Sect. 2.2. Then, we provide an overview of some technical details related to WSN standards and communication technologies for CH monitoring in Sect. 2.3. Section 2.4 presents an architecture of a WSN for monitoring CH objects. The important design factors regarding WSN-based monitoring of the CH are discussed in Sect. 2.5. WSN deployment aspects and its requirements for monitoring the CH objects are discussed in Sects. 2.6 and 2.7, respectively. Section 2.8 gives future research directions and discusses some open research issues. And, finally, Sect. 2.9 concludes this chapter.

2.2 Related Works

In recent decade, WSNs have been extensively utilized for monitoring the CH objects. This section reviews some of these recent studies. In [9], the authors used a WSN to monitor the health state of a heritage building in real time by measuring a number of important parameters such as humidity, temperature, masonry cracks, and visual light. Similarly, the study in [10] provides a remote assessment of CH environments using wireless sensor array networks. The system consists of an array of piezoelectric quartz crystals covered with various metals such as iron and copper, and contains a temperature and a humidity sensor. The system uses IEEE 802.15.4 low-power radio as the communication module. The measurement results can be sent to an Internet server remotely for real-time visualization.

In [11], a system is proposed to remotely monitor an archeological site by measuring the temperature and atmospheric pressure in the site. The system utilizes an electronic card based on microcontrollers and transfers the measured data to a unit supported by Raspberry Pi (RPi) board which is able to take pictures of the site using a high-quality camera connected to it. In addition, the proposed system uses ZigBee communication technology for local data transmission, and satellite and WiMAX technologies for the communication with the remote server.

The authors in [12] propose an effective CH object tracking and deformation detection technique using WSNs in order to monitor and detect deformation of the CH objects in early stages. They discover a connected core to make a backbone route for collection and transmission of messages among the sensor nodes which consequently results in the reduction of energy consumption and communication costs. The analysis and results of the proposed method reveal that it performs better compared to the existing methods in terms of the accuracy of the deformation detection as well as the network traffic.

Moreover, recent advances on the field of WSNs have significantly contributed to employment of structural health monitoring (SHM) related to the CH constructions. For example, in [13], temperature and humidity sensors are used to monitor the environment in a medieval tower. The SHM provides information on the displacement and vibration of the tower. In addition, the study shows how the obtained data can be utilized to predict the unusual behaviors of the tower under study. In [14], a WSN consisting of several temperature and fiber optical displacement sensors is used to monitor the damages on a historical site in Portugal. The obtained data is used to update a finite element model (FEM) in order to assess the damage state of the site. Similarly, the authors in [15] describe the strategies of a WSN design and data processing in a church focusing on seismic assessment. The SHM focuses on the management of the safety of the church in various situations where natural disasters such as earthquake may happen. The data acquisition system is wirelessly connected to a signal-processing unit and the collected data is used to update the FEM on the church.

The project in [16] deploys a WSN to monitor the health of artworks in museums by considering various critical environmental parameters. The proposed system contains several tools for real-time alerting and network management, as well as data visualization in the environment. However, these functionalities are not embedded together and the system is not free. The project in [7] integrates all the functionalities of the WSN available in [16], and introduces the functionality of indoor geospatial data layering too. In addition, the system is freely available.

This chapter attempts to present a comprehensive review on the use of WSNs in monitoring the CH objects. Therefore, the contributions of this study relative to the current literature can be summarized as follows:

- To the best of our knowledge, this is the first survey study that provides a comprehensive review on the use of WSNs in monitoring the CH objects.
- We provide a classification for critical design factors of the WSN-based monitoring of the CH objects.
- We also discuss different WSN deployment strategies and provide a summary of the important requirements for the design and implementation of the WSNs in the CH context.
- Finally, we provide future research direction and discuss some open research issues.

2.3   WSN Standards and Communication Technologies for CH onitoring

WSN protocols develop cost-effective and standard-based solutions supporting low power consumption and low data rates. Examples of communication protocols that can be used for monitoring of CH objects in the IoT era are Bluetooth, IEEE 802.15.4, ZigBee, and Wi-Fi. Some specific standards designed for industrial applications are also in use such as WirelessHART and ISA100.11a. In this section, we briefly discuss these standards and communication protocols.

2.3.1 Bluetooth

Bluetooth is a technology that is mainly utilized for the communication between wireless devices within short distances. It operates in the 2.4 GHz frequency band (between 2400 and 2483.5 MHz) [17] and targets the personal area networks. Compared to other communication standards, Bluetooth devices have limited battery lifetime. In addition, the communication range for Bluetooth devices is around 10 m, which in turn limits the deployment of this technology in wide areas [18].

2.3.2 IEEE 802.15.4

IEEE 802.15.4 is a low-power and low-cost wireless standard which defines the operation for physical and data link layers of the low-rate wireless personal area networks (LR-WPANs). It provides connectivity within short ranges (e.g., up to 20 m) making it appropriate for the use in WSNs [19].

2.3.3 ZigBee

ZigBee is a communication technology based on IEEE 802.15.4 standards. It provides low power consumption and low data rate communication [20], and is created and designed for wireless controls and sensors. ZigBee provides connectivity within a range of normally 50 meters and its data rate is 250 kbps at 2.4 GHz, 40 kbps at 915 MHz, and 20 kbps at 868 MHz [17].

2.3.4 Wi-Fi

Wi-Fi is a technology suitable for communication within a range of up to 100 meters [21] which can be used for WSNs and communication of IoT devices. Wi-Fi-based WSNs used in CH monitoring have the characteristics of high data rate and bandwidth, and large-scale data collection, as well as the capability of video monitoring of CH objects which may not be realized using ZigBee technology [22].

2.3.5 WirelessHART

WirelessHART is an industrial control protocol and an extension of the Highway Addressable Remote Transducer (HART) communication protocol which is designed to be reliable, easy to use, and interoperable protocol deployed in process control applications. WirelessHART has low power consumption compared to ZigBee with higher security standards [23]. In addition, it is a deployed protocol for industrial WSNs which reduces the cost of automation systems and plays an important role in alarm management systems using real-time data transfer.

2.3.6   ISA100.11a

ISA100.11a is another communication protocol specially designed for industrial WSNs. It has the features of low power consumption, reliability, scalability, and security as well as high real-time data transfer in industrial environments [24]. ISA 100.11a operates on 2.4 GHz frequency band and supports high data rates up to 250 kpbs. Similar to WirelessHART, ISA100.11a is an industrial control protocol deployed in process control applications.

2.4   Network Architecture

In this section, an architecture of a WSN for monitoring CH objects is presented. This architecture, as shown in Fig. 2.1, consists of three layers: sensor layer, gateway layer, and database layer. We briefly discuss these layers and their functions in the following subsections.

2.4.1   Sensor Layer

In the sensor layer, various sensors are deployed to sense, measure, and gather information in the environment.

The deployed sensors periodically sense the required environmental parameters such temperature, humidity, and light. Then, they transfer data to the gateway layer using a radio module for wireless communication such as ZigBee or Wi-Fi.

2.4.2   Gateway Layer

This layer has the local information processing capabilities. It receives data from the sensors deployed in the sensor layer. Moreover, it has the capabilities of analyzing data, and, in case of any abnormalities, it can send an alarm message to the responsible authority. In addition, it can communicate with the database layer to store the data obtained from the sensor nodes.

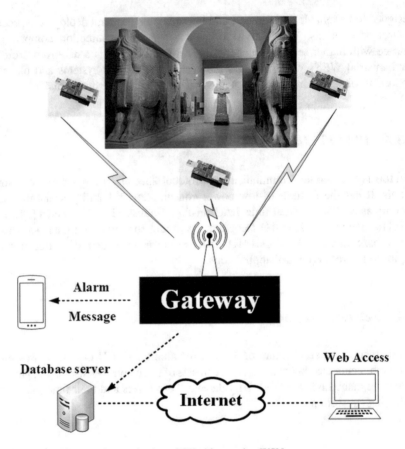

Fig. 2.1 An architecture for monitoring of CH objects using WSNs

2.4.3 Database Layer

The database layer is capable of remote monitoring and storage at any time and any place. It can store the sensed data in the database, and administrators can have access to the environmental information using a web interface. Therefore, the CH objects can be monitored in real time and required actions can be taken accordingly.

2.5 Design Factors of the WSN-Based Monitoring of the CH

Monitoring CH using WSNs requires special attention to make sure an efficient and reliable system, and acceptable quality of service (QoS). The important design factors regarding WSN-based monitoring of CH are briefly discussed in this section.

Table 2.1 A summary of the design factors for monitoring of the CH utilizing WSNs

Ref	Primary design factors				Secondary design factors			
	EE	QoS	CC	Accuracy	Security	Cost	Fault tolerance	Reliability
[3]	X	–	X	X	–	X	–	X
[4]	X	X	–	X	–	X	–	–
[6]	X	X	–	X	–	X	–	–
[7]	X	–	–	X	–	X	–	X
[8]	X	–	X	–	–	X	–	X
[9]	X	–	–	X	–	-	–	X
[10]	X	–	X	X	–	–	–	–
[11]	X	X	–	X	X	X	X	X
[12]	X	–	X	X	–	X	–	–
[13]	X	–	X	X	–	X	–	X
[14]	-	–	–	X	–	X	–	X
[15]	X	–	X	–	X	X	–	X
[16]	X	–	X	X	X	–	–	–

EE energy efficiency, *QoS* quality of service, *CC* coverage and connectivity
– = not considered; *X* = considered

We classify these factors into primary vs. secondary design factors. Primary design factors are the most important factors of the targeted system in order to achieve a reliable and efficient system, and the secondary design factors are those which are less important compared to the primary design factors but still can significantly affect the overall performance of the system. A summary of these design factors for WSN-based monitoring of the CH is presented in Table 2.1.

2.5.1   Primary Design Factors

The primary design factors for the WSN-based monitoring of the CH are those factors having the utmost importance towards achieving an efficient and reliable system.

Energy Efficiency (EE)

WSNs are usually restricted in power resources. They should operate under the limitations of low power transmission and low energy consumption. Therefore, deploying WSN-based systems that are energy-efficient is a need for monitoring CH in order to maximize the energy saving of the system.

QoS

QoS is one of the most important factors in wireless sensor applications. It is difficult and challenging to guarantee QoS in WSNs because the available resources and different applications running over the network have various constraints in their requirements and nature. Therefore, it is important to consider appropriate metrics of QoS for WSNs especially in monitoring CH objects in order to provide reliability and availability towards achieving high QoS.

Coverage and Connectivity (CC)

Coverage and connectivity are other important factors in monitoring CH objects for the communication of sensory data from the sensor nodes to the sink. There should be an efficient algorithm to calculate the minimum number of required sensors to fully cover the area under study in order to guarantee the desired QoS in WSNs.

Accuracy

It is important to design accurate WSN-based models for monitoring CH objects to make sure that the models can work properly in various situations such as disaster in order to improve the performance of the models. Accurate WSNs can efficiently utilize wireless resources and guarantee an efficient operation in real time.

2.5.2 Secondary Design Factors

The secondary design factors for the WSN-based monitoring of the CH are those factors which have less importance compared to primary ones but still can significantly affect the performance of the system under study. These factors are briefly discussed in the following subsections.

Security

Security in WSNs has always been a main concern in order to have a secure communication between the sensor nodes. Due to the limitation of sensor nodes in terms of battery and available bandwidth for communication, it would be more challenging to provide security for WSNs compared to the traditional networks. In the case of CH objects, lack of secure communication may cause security attacks by the attackers, which in turn ruins the CH objects. Therefore, it is important to have a secure communication between the sensor nodes in order to detect any malicious activities, and ensure the accuracy of routing stats in monitoring CH objects as invaluable historical legacy.

Cost

In WSNs, a high-cost sensor node includes high computational power, which in turn increases the energy consumption of the entire system. In order to have a cost-effective WSN, and depending on the monitoring parameters for the CH objects, the WSN should consist of both low-cost sensor nodes to perform simple tasks and high-cost sensor nodes to execute complex tasks.

Fault Tolerance

Fault tolerance is another important design factor in WSN-based monitoring of the CH objects. A WSN may be faulty due to different reasons such as human interference, hardware issues, and various environmental factors. In order to have an efficient WSN, it should be able to tolerate the aforementioned failures. The most common technique to handle such failures is to deploy and utilize extra sensor nodes as backup nodes during faulty time [25].

Reliability

Reliability is another factor in resource-constraint WSNs in order to achieve efficient control and monitoring systems. Fast response is needed for WSN-based monitoring of CH objects to take the necessary actions on time, especially in disaster situations. For example, the WSN should decide very fast about the next hop for data routing in the network to ensure that the communication is done on time.

2.6 WSNs Deployment Aspects in CH

In order to monitor the CH objects using WSNs, it is important to determine the right location of the sensor nodes and employ proper algorithms and strategies for the deployment issues. In this regard, deployment strategies can be classified into random vs. deterministic deployment.

2.6.1 Random Deployment

In random deployment, sensor nodes are randomly placed in the environment to decrease the cost of deployment. They can also be placed where the density of the distributed sensor nodes is not uniform in the area. For example, in [26], the authors use random deployment strategy for deployment of relay nodes in a 2D plane in order to propose a network lifetime maximization algorithm when there is a direct communication between the relay nodes and the base station (BS). The study in [12] utilizes a random strategy for deployment of sensor nodes to provide an effective CH object tracking and deformation detection technique using WSNs. This technique monitors and detects deformation of the CH objects in early stages. The authors discover a connected core to provide a backbone route for collection and transmission of messages among the sensor nodes. This in turn results in the reduction of energy consumption and communication costs. In addition, the authors in [27] consider random deployment of sensor nodes in the WSNs for monitoring of the CH sites which are typically placed on the 3D surface of a constrained space [28]. The sensor nodes are randomly deployed in the area to provide full coverage of a given set of target objects with a minimal number of sensors and high level of sensing reliability in order to ensure the long-term conservation.

2.6.2   Deterministic Deployment

Unlike random deployment, sensor nodes in deterministic deployment are placed in specific predefined locations. This type of deployment is also called grid-based deployment. Deterministic or grid-based deployment can significantly improve the network lifetime and fault-tolerance issues as well as the network connectivity if it is coupled with multipath routing [29]. The deterministic deployment can be divided into two categories: static vs. dynamic [30]. In static deployment, the sensor nodes cannot change their location after deployment. If the sensor elements can change their positions, it is called dynamic deployment of sensor nodes. This type of deployment can also be divided into regular (e.g., hexagon, square, and triangle) and irregular patterns [31].

In [32], the authors propose an integer linear programming (ILP) algorithm to find the optimal placement configuration of the sensor nodes using deterministic deployment approach. The approach reveals the necessary parameters required to reduce the number of sensors for monitoring the area of interest and consequently reduce the final cost of the WSN. These parameters include the number and location of the BSs, orientation and field of view, as well as the sensing range. Similarly, the study in [33] proposes a solution for optimal coverage of the sensors when they are precisely located within the monitored area. Moreover, the approach finds the optimal number of sensor nodes to be deployed using a deterministic deployment approach by considering the coverage density and the relation between the sensor nodes and the covered area.

In [34], the authors propose a deterministic deployment algorithm in order to optimize the number of sensors required to cover the area of interest and to determine their locations. The approach utilizes a probabilistic optimization algorithm to particularly model the specific issues related to the coverage of grid points based on the security and their tactical importance. In addition, the study in [35] considers a grid-based deployment of WSNs with nodes located in a square of unit area with the focus on the connectivity of the network as well as on the coverage of the monitored area. Moreover, the authors in [36] propose a centralized and deterministic approach to remove the coverage holes close to the boarder of sensing environment and the obstacles. The Delaunay triangulation (DT) deployment method is then applied for the uncovered areas. Particularly, before a sensor is deployed each candidate position is scored according to the probabilistic sensor detection model, which in turn provides the optimal location of the new sensors in terms of coverage gains.

2.7   Requirements for WSNs-Based Monitoring of CH

After careful study on the existing literature in monitoring environmental systems and CH objects, it is crucial to take into consideration a number of requirements before the design and the implementation of a WSN-based solution. These requirements are discussed in this section.

- **Low Deployment Cost:** Depending on the size and area of the CH objects, thousands of nodes may be required to be deployed in the area to provide wider coverage. Therefore, the sensor nodes should be easy to deploy and should have a low cost as well since available WSN platforms in the market are quite expensive. Thus, it is required to have cheaper WSN-based platform for monitoring the CH objects.
- **Energy-Efficient Sensor Nodes:** It is significantly important to have power-efficient sensor nodes to monitor the CH objects especially for long-term environmental monitoring. In this regard, clustering methods [37], data aggregation [38], and cross layer protocols [39] are proposed as the potential solutions to enhance the lifetime of the sensor nodes.
- **Scalability:** As discussed in previous sections, there may be thousands of sensor nodes deployed in the environment for monitoring the CH objects. Therefore, it is important to prove that the theoretical solutions available are applicable for real deployment of the WSN.
- **Remote Access:** It is another important requirement of the WSN-based monitoring of CH objects as the data available from the sensor nodes may be required to be available to responsible people in various geographical areas. In this regard, various communication protocols such as ZigBee and WirelessHART can be used to send the data to the gateway where the sensed data are stored.

2.8   Open Research Issues

Despite different research activities carried out and the recent development that has been achieved, there are still several issues and challenges that need to be carefully considered regarding the design and deployment of WSNs in monitoring the CH objects. Some of the most important challenges and research issues are discussed in this section.

2.8.1 Security and Privacy

WSN-based monitoring of the CH objects provides a great opportunity of interaction between the sensors and the environment itself. However, the ad hoc nature and the wireless vulnerability make it a great target for security attacks. Attackers can alter the sensitive data collected by the sensors and disturb the normal activates of the network which may jeopardize the health of the CH object. Therefore, it is crucial to design and implement a security-aware WSN mechanism for monitoring the CH objects.

2.8.2 Coverage and Connectivity

In WSNs, it is important to manage energy resources and provide reliable QoS. In this regard, network coverage and connectivity would be important factors to consider. Node deployment strategies should reduce the communication overhead and minimize cost while providing a high degree of coverage and maintaining a connected network simultaneously. The coverage requirements may vary according to the object. For example, a CH object inside a building possibly requires a low degree of coverage. On the other hand, monitoring a large CH building possibly requires a high degree of coverage which is done by multiple nodes. Thus, it is important to have the optimal number of nodes in the environment while providing full coverage to the object. Therefore, it is necessary to define and consider precise measures of the network coverage and connectivity that affect the overall performance of the system.

2.8.3 Power Consumption

Power consumption is a critical issue for long-term operations. Environmental monitoring systems are generally considered as high power consumption systems since various sensors such as gas sensors are considered as energy hungry devices [40]. Therefore, it is important to have an energy-efficient WSN for monitoring the CH objects. This can be achieved by using efficient communication protocols such as Bluetooth Low Energy (BLE) [41] as well as solutions based on LoRa and SigFox technologies [42].

2.8.4 Scalability

Sensor nodes deployed in WSNs should be scalable and easy to deploy and maintain in large environments. For example, systems based on Bluetooth and General Packet Radio Service (GPRS) technologies face the problem of low scalability [43].

Therefore, it is beneficial to investigate into communication technologies in order to provide a scalable WSN which allows the increase in the number of sensors as well as the integration of new services in the system.

2.8.5   Cost

Cost is an important factor when it is required to deploy a large number of sensors for monitoring the CH objects. A number of studies such as [44, 45] use GPRS as the main communication technology among sensor nodes which results in a higher cost per node. In addition, some systems depend on a localization method, which increases the overall cost of the system [46]. Therefore, it is important to conduct research into the communication technologies and localization techniques for WSN-based monitoring of the CH objects in order to design and implement a cost-effective system.

2.9   Conclusion

The integration of WSNs and the IoT can be a promising solution regarding monitoring of the CH objects and improving their security-related issues in smart environments. In this chapter, through a comprehensive investigation on the use of WSNs in the CH context, we discussed the communication protocols that can be used for monitoring of the CH objects in the IoT era. Moreover, we presented a network architecture and discussed the important design factors regarding the WSN-based monitoring of the CH objects in the IoT era in order to ensure an efficient and reliable monitoring system. We also discussed the WSN deployment aspects and presented the most important requirements that need to be taken into account before the design and implementation of a reliable solution in the CH context. Finally, we outlined some research challenges and open issues that need to be carefully taken into consideration in this area.

References

1. E. Ever, F.M. Al-Turjman, H. Zahmatkesh, M. Riza, Modelling green HetNets in dynamic ultra large-scale applications: A case-study for femtocells in smart-cities. Comput. Netw. **128**, 78–93 (2017)
2. A. Al-Fuqaha, M. Guizani, M. Mohammadi, M. Aledhari, M. Ayyash, Internet of things: A survey on enabling technologies, protocols, and applications. IEEE Commun. Surv. Tutor. **17**(4), 2347–2376 (2015)
3. F. Mesas-Carrascosa, D. Verdú Santano, J. Meroño de Larriva, R. Ortíz Cordero, R. Hidalgo Fernández, A. García-Ferrer, Monitoring heritage buildings with open source hardware sensors: A case study of the mosque-cathedral of Córdoba. Sensors **16**(10), 1620 (2016)

4. M.A. Rodriguez-Hernandez, Z. Jiang, A. Gomez-Sacristan, and V. Pla, "Intelligent municipal heritage management service in a smart city: Telecommunication traffic characterization and quality of service", Wirel. Commun. Mob. Comput. (2019)
5. D. Camuffo, *Microclimate for Cultural Heritage: Conservation, Restoration, and Maintenance of Indoor and Outdoor Monuments* (Elsevier, Amsterdam, 2013)
6. L. D'Alvia, E. Palermo, Z. Del Prete, Validation and application of a novel solution for environmental monitoring: A three months study at "Minerva Medica" archaeological site in Rome. Measurement **129**, 31–36 (2018)
7. F. D'Amato, P. Gamba, E. Goldoni, *Monitoring Heritage Buildings and Artworks with Wireless Sensor Networks*, In IEEE Workshop on Environmental Energy and Structural Monitoring Systems (EESMS) (2012), pp. 1–6
8. M. Ceriotti, L. Mottola, G.P. Picco, A.L. Murphy, S. Guna, M. Corra, M. Pozzi, D. Zonta, P. Zanon, *Monitoring Heritage Buildings with Wireless Sensor Networks: The Torre Aquila Deployment*, In Proceedings of the IEEE International Conference on Information Processing in Sensor Networks (2009), pp. 277–288
9. A. Mecocci, A. Abrardo, Monitoring architectural heritage by wireless sensors networks: San Gimignano—A case study. Sensors **14**(1), 770–778 (2014)
10. H. Agbota, J. Mitchell, M. Odlyha, M. Strlič, Remote assessment of cultural heritage environments with wireless sensor array networks. Sensors **14**(5), 8779–8793 (2014)
11. F. Leccese, M. Cagnetti, A. Calogero, D. Trinca, S. Pasquale, S. Giarnetti, L. Cozzella, A new acquisition and imaging system for environmental measurements: An experience on the Italian cultural heritage. Sensors **14**(5), 9290–9312 (2014)
12. Z. Xie, G. Huang, R. Zarei, J. He, Y. Zhang, H. Ye, Wireless sensor networks for heritage object deformation detection and tracking algorithm. Sensors **14**(11), 20562–20588 (2014)
13. D. Zonta, H. Wu, M. Pozzi, P. Zanon, M. Ceriotti, L. Mottola, G.P. Picco, A.L. Murphy, S. Guna, M. Corrá, Wireless sensor networks for permanent health monitoring of historic buildings. Smart Struct. Syst. **6**(5–6), 595–618 (2010)
14. H.F. Lima, R. da Silva Vicente, R.N. Nogueira, I. Abe, P.S. de Brito Andre, C. Fernandes, H. Rodrigues, H. Varum, H.J. Kalinowski, A. Costa, J. de Lemos Pinto, Structural health monitoring of the church of Santa casa da Misericórdia of Aveiro using FBG sensors. IEEE Sensors J. **8**(7), 1236–1242 (2008)
15. F. Potenza, F. Federici, M. Lepidi, V. Gattulli, F. Graziosi, A. Colarieti, Long-term structural monitoring of the damaged basilica S. Maria di Collemaggio through a low-cost wireless sensor network. J. Civ. Struct. Heal. Monit. **5**(5), 655–676 (2015)
16. L.M.P.L. de Brito, L.M.R. Peralta, F.E.S. Santos, R.P.R. Fernandes, *Wireless Sensor Networks Applied to Museums' Environmental Monitoring*, In The Fourth International Conference on Wireless and Mobile Communications (2008), pp. 364–369
17. T. Alhmiedat, A survey on environmental monitoring systems using wireless sensor networks. JNW **10**(11), 606–615 (2015)
18. N. Baker, ZigBee and Bluetooth: Strengths and weaknesses for industrial applications. Comput. Control. Eng. **16**(2), 20–25 (2005)
19. F. Al-Turjman, E. Ever, H. Zahmatkesh, Small cells in the forthcoming 5G/IoT: Traffic modelling and deployment overview. IEEE Commun. Surv. Tutor. **21**(1), 28–65 (2018)
20. P. Kinney, Zigbee technology: Wireless control that simply works, In *Communications Design Conference*, Vol. 2 (2003), pp. 1–7
21. E. Ferro, F. Potorti, Bluetooth and Wi-fi wireless protocols: A survey and a comparison. IEEE Wirel. Commun. **12**(1), 12–26 (2005)
22. L. Li, H. Xiaoguang, C. Ke, H. Ketai, *The Applications of Wifi-Based Wireless Sensor Network in Internet of Things and Smart Grid*, In 6th IEEE Conference on Industrial Electronics and Applications (2011), pp. 789–793
23. P. Ferrari, A. Flammini, S. Rinaldi, E. Sisinni, *Performance Assessment of a WirelessHART Network in a Real-World Testbed*, In IEEE International in Instrumentation and Measurement Technology Conference (MTC) (2012), pp. 953–957
24. ISA100, Wireless Systems for Automation, https://www.isa.org/isa100/

25. A. Tripathi, H.P. Gupta, T. Dutta, R. Mishra, K.K. Shukla, S. Jit, Coverage and connectivity in WSNs: A survey, research issues and challenges. IEEE Access **6**, 26971–26992 (2018)
26. K. Xu, H. Hassanein, G. Takahara, Q. Wang, Relay node deployment strategies in heterogeneous wireless sensor networks. IEEE Trans. Mob. Comput. **9**(2), 145–159 (2009)
27. C. Wu, L. Wang, On efficient deployment of wireless sensors for coverage and connectivity in constrained 3D space. Sensors **17**(10), 2304 (2017)
28. J. Joo, J. Yim, C.K. Lee, Protecting cultural heritage tourism sites with the ubiquitous sensor network. J. Sustain. Tour. **17**(3), 397–406 (2009)
29. F. Al-Turjman, *Wireless Sensor Networks: Deployment Strategies for Outdoor Monitoring* (CRC Press, Boca Raton, 2018)
30. D. Costa, L.A. Guedes, The coverage problem in video-based wireless sensor networks: A survey. Sensors **10**(9), 8215–8247 (2010)
31. H.P. Gupta, P.K. Tyagi, M.P. Singh, Regular node deployment for k-coverage in m-connected wireless networks. IEEE Sensors J. **15**(12), 7126–7134 (2015)
32. Y.E. Osais, M. St-Hilaire, F.R. Yu, Directional sensor placement with optimal sensing range, field of view and orientation. Mob. Net. Appl. **15**(2), 216–225 (2010)
33. X. Han, X. Cao, E.L. Lloyd, C.C. Shen, *Deploying Directional Sensor Networks with Guaranteed Connectivity and Coverage*, In 5th Annual IEEE Communications Society Conference on Sensor, Mesh and Ad Hoc Communications and Networks (2008), pp. 153–160
34. S.S. Dhillon, K. Chakrabarty, *Sensor Placement for Effective Coverage and Surveillance in Distributed Sensor Networks*, In IEEE Wireless Communications and Networking (WCNC), Vol. 3 (2003), pp. 1609–1614
35. S. Shakkottai, R. Srikant, N.B. Shroff, Unreliable sensor grids: Coverage, connectivity and diameter. Ad Hoc Netw. **3**(6), 702–716 (2005)
36. C.H. Wu, K.C. Lee, Y.C. Chung, A Delaunay triangulation based method for wireless sensor network deployment. Comput. Commun. **30**(14–15), 2744–2752 (2007)
37. A. Kumar, *Energy Efficient Clustering Algorithm for Wireless Sensor Network, Doctoral Dissertation* (Lovely Professional University, Punjab, 2017)
38. A. Taherpour, M. Chobin, M. Rahmani, *Collaborative Data Aggregation Using Multiple Antennas Sensors and Fusion Center with Energy Harvesting Capability in WSN* (IET Communications, 2019)
39. H. Hadadian, Y.S. Kavian, *Cross-Layer Protocol Using Contention Mechanism for Supporting Big Data in Wireless Sensor Network*, In 10th IEEE International Symposium on Communication Systems, Networks and Digital Signal Processing (CSNDSP) (2016), pp. 1–5
40. V. Jelicic, M. Magno, D. Brunelli, G. Paci, L. Benini, Context-adaptive multimodal wireless sensor network for energy-efficient gas monitoring. IEEE Sensors J. **13**(1), 328–338 (2012)
41. K. Sornalatha, V.R. Kavitha, *IoT Based Smart Museum Using Bluetooth Low Energy*, In 3rd IEEE International Conference on Advances in Electrical, Electronics, Information, Communication and Bio-informatics (AEEICB) (2017), pp. 520–523
42. A. Perles, E. Perez-Marin, R. Mercado, J.D. Segrelles, I. Blanquer, M. Zarzo, F.J. Garcia-Diego, An energy-efficient internet of things (IoT) architecture for preventive conservation of cultural heritage. Futur. Gener. Comput. Syst. **81**, 566–581 (2018)
43. O.A. Postolache, J.D. Pereira, P.S. Girao, Smart sensors network for air quality monitoring applications. IEEE Trans. Instrum. Meas. **58**(9), 3253–3262 (2009)
44. A. Kadri, E. Yaacoub, M. Mushtaha, A. Abu-Dayya, *Wireless Sensor Network for Real-Time Air Pollution Monitoring*, In 1st IEEE International Conference on Communications, Signal Processing, and their Applications (ICCSPA) (2013), pp. 1–5
45. A.R. Al-Ali, I. Zualkernan, F. Aloul, A mobile GPRS-sensors array for air pollution monitoring. IEEE Sensors J. **10**(10), 1666–1671 (2010)
46. M. Navarro, T.W. Davis, Y. Liang, X. Liang, *A Study of Long-Term WSN Deployment for Environmental Monitoring*, In 24th IEEE Annual International Symposium on Personal, Indoor, and Mobile Radio Communications (PIMRC) (2013), pp. 2093–2097

Chapter 3
Tracking and Analyzing Processes in Smart Production

Selver Softic, Egon Lüftenegger, and Ioan Turcin

3.1 Introduction

The rise of the Industry 4.0 opens a new era of innovation and change. Internet of Things (IoT) and other enabler technologies such as Big Data will become decisive factors of future success for companies and global industrial systems [1]. In this context the business models and related processes are main success factors for innovations [2] and they are essential for strategic management decisions of the companies. Especially manufacturing oriented enterprises will face the challenge to develop innovative technology driven business models alongside technology innovations [1]. In this field, this will be essential action for securing the future competitiveness. Failing to develop technology driven business models around industrial Internet innovations in an internationally highly competitive environment will have serious implications, on companies and their surrounding. To identify risks and challenges and to better understand what paradigm shift to Smart Production means when it comes to business model we need reliable tools that can support this process. Although there are automated tools for business process monitoring, there is a lack of management tools that guide the process of capturing KPIs (Key Performance Indicators). In particular, this gap was noticeable in Smart Production scenarios driven by Industry 4.0. Smart Production integrates business processes, humans, and technologies into different manufacturing scenarios. In these scenarios we encounter tasks and activities covering the following three dimensions: Humans, Systems, and Machines. By taking those dimensions into account, we can observe

S. Softic (✉) · E. Lüftenegger
IT and Business Informatics, CAMPUS 02 University of Applied Sciences, Graz, Austria
e-mail: selver.softic@campus02.at; egon.lueftenegger@campus02.at

I. Turcin
Automation Technology, CAMPUS 02 University of Applied Sciences, Graz, Austria
e-mail: ioan.turcin@campus02.at

© Springer Nature Switzerland AG 2020
F. Al-Turjman (ed.), *Trends in Cloud-based IoT*, EAI/Springer Innovations
in Communication and Computing, https://doi.org/10.1007/978-3-030-40037-8_3

their impact and relevance for a Smart Production process. Therefore, there is a plenty of potential for optimization of such processes. In order to optimize such processes, an important pre-step represents the monitoring and tracking of activities, tasks, and actors involved. With this procedure we are able to uncover the weaknesses in process handling. Further, we can perform process-driven analysis of business and production data, e.g., obtained from sensors. Based on such insights it allows us to run benchmarking based on measurable process indicators and use past activities to drive perspective interaction with the customers. For this purpose, we developed a management tool. The developed tool serves as instrument that supports the business process/manufacturing monitoring through the visual attachment and tracking of KPIs for tasks and activities. The aim of applying the management tool is that companies can evaluate to what extent Smart Production processes are implemented and, based on that, define a strategy or increasing performance. Thus, in order to demonstrate the usefulness of our software we will apply the tool on a generalized production scenario from literature and draw some preliminary findings from our experiment. Additionally we will also present an extended scenario including IoT technologies.

3.2   Related Work

3.2.1   BPMN and Balance Score Card

BPMN is de facto standard for business process specification [3]. The ability to support BPMN is from high relevance for process monitoring applications. The process and collaboration diagram that was created with BPMN is one of the most frequently used and therefore most important forms of business process model representations [4]. Regarding Smart Production, there are modeling approaches that are extending BPMN for modeling cyber-physical systems [5] or even implementing a whole new notation [6]. However, BPMN lacks the expressiveness for KPIs tracking. There are commercial tools for KPI performance monitoring but those tools are designed for setting thresholds that are monitored by a Business Process Modeling (BPM) engine. They do not let users set as-is and to-be values. Hence, they are only good at operational level, but not at strategic level. At strategic level, there is a need for tracking and measuring the change or the desire of changing. At operational level, the focus lies on the results of a process execution. Thus, implemented process management tool offers a wide range of applications for companies because it relies on BPMN standard and tries additionally to fill the identified missing gaps in performance tracking through extending the principles and benefits of Balanced Scorecard (BSC) [7].

3.2.2 KPIs and KPI Tracking

KPI tracking refers to all the tools and methods that companies use to monitor their performance metrics. When measuring key performance indicators (KPIs) data is collected and converted into useful metrics that are measured and displayed in digestible charts and dashboards. The KPIs provide insights into different aspects of processes and can vary depending on requirements, industries, and even specific departments. However, KPI tracking is not just about collecting data, but also to place these metrics in a wider context in order to determine whether they indicate a degree of improvement or whether it is necessary to optimize certain areas of an organization. A key goal of tracking KPIs is to measure progress towards specific goals and long-term goals or the quantifiable milestones. In this sense, KPI tracking serves as a benchmark for progress and improvement.

3.2.3 Enabling Technologies for Smart Production

In following subsection we give a short overview of related work describing significant aspects of enabling technologies and concepts for Smart Production.

Internet of Things (IoT) and Cloud Computing

The basic concept behind Internet of Things (IoT) is the concept of connecting any uniquely identifiable device to the Internet and to other connected devices. Meanwhile, Internet of Things (IoT) influences our everyday life and cloud computing plays a significant role as an infrastructural realization of it. The IoT allows people and things (e.g., sensors, actuators, and smart devices) to be connected ubiquitously and any time. IoT is also one of the enabling technologies for Smart Production, and its application is dramatically growing, same as demand for reliable cloud-based applications that rely on well-scheduled and performing algorithms for sensor networks [8]. The application domains of IoT are manifold starting by, i.e., managing city infrastructures like parking lots in a smart way [9], over tackling the security issues in 5G networks [10] up to managing smart energy grids [11].

Big Data

Beside IoT a significant enabling technology for Smart Production is Big Data. In the context of the Smart Factory approaches using artificial intelligence methods to predict, calculate, or track the indicators for Big Data based services will play an decisive role in the future [12, 13]. With the explosion of global data, the term Big Data is mainly used to describe huge data sets [14]. Essentially, data is characterized by three properties, which, according to their English names, are referred to as the "three V's" of Big Data: Volume, Velocity, and Variety. While Volume describes the ever-increasing volume of data, Velocity refers to the speed of traffic, and Variety refers to the nature of data occurring in multiple and differently structured data sources. Very often, the term Value as the fourth "V" has been added to this concept. An (extended) benefit can arise when large amounts of data are analyzed in order to identify an economic exploitation potential [15, 16]. Thus, Big Data opens up new opportunities to discover new utility values and to better understand already existing but hidden values [15]. Compared to traditional data sets, Big Data typically contains masses of unstructured data that lead to new challenges such as the effective organization and management of such data sets and as well as enabling real-time analytics [14]. Big Data also represents a paradigm shift in the way data is analyzed. Data is now "made to talk" through correlation analysis and pattern identifications, which means that data scientists as experts in computer-aided data analysis can derive new knowledge from Big Data. Nowadays, far more data can be analyzed than ever before, a circumstance that makes new ways of gaining knowledge feasible. The huge amount of data makes it possible to reveal complex relationships through skillful questions to the data [15].

Smart Factory and Cyber-Physical Systems

Industry 4.0 describes the fourth industrial revolution, driven by information and communication technologies (ICT) and the Internet of Things (IoT) [17]. Internationally, Industry 4.0 is the digitization of industry, with the goal of horizontal and vertical integration of the value chains, whereby both the processes along the value chain with suppliers and customers and the communication between man, machine, and resources should be completely automated. The idea of autonomously controlling and optimizing production systems as well as intelligent workpieces is an integral part of Industry 4.0 or Smart Factory. The technological prerequisites for this are the so-called cyber-physical systems (CPS). CPS are considered as the connection of uniquely identifiable physical "things" or objects to the Internet or other comparable virtual structure what corresponds to the IoT paradigm. The production processes are not only actively controlled by networked CPS, but sometimes also offer platforms for innovative business models and services. Thus, not only does the corporate strategy have to be adapted, but also the existing "mono-organizational" business models have to be completely re-thought and redefined [18–20]. One of the main benefits of the IoT is also the ability to use digitally networked products, services, and solutions as manufacturers try to deepen the role of the customer into the value creation process [21]. The networking of intelligent products increases the quality of automation and thereby increases the competitiveness of high-wage locations [22]. Due to the networking of production systems and the resulting increase in machine-to-machine communication (M2M), not only the number of sensors required and installed in the system is increased, but also the overall complexity of the system in general affected. The accumulation of sensors and thus increasing overall complexity as a result of ongoing digitization leads lately to generation of data in large quantities, in this context also known as Big Data.

3.3 Methodology

As overall methodology we use "Design Science." The "Design Science" is suited for the evaluation of conceptual artifacts such as methods and software [23]. We use the design science approach to validate our software-supported method with a case study [24]. The first step of applied methodology foresees the identification of key figures which were already determined by the respective company in this context. For this purpose, the activities and sub-processes are examined individually. In this step, the method tries to identify which stakeholders are affected and what KPIs are already collected by them or others. The identified KPIs are recorded, assigned to the dimensions of the Balanced Scorecard (BSC) [7], and visualized. The first measurement gaps in the process as well as a possible imbalance in the BSC are already visible here. Both circumstances were discussed in detail in a second step. In the course of an open brainstorming, it is first possible to discuss and cover more global needs (which cannot be assigned to individual gaps). Then, using the prepared performance index (consisting from previously identified and assigned KPIs), you can search for performance indicators that close the measurement gaps in a targeted manner. Finally, in the last step target values for all KPIs are to be defined. Enabling the business process performance tracking in Smart Production through application of proposed management KPI tracking tool requires beside the appropriate modeling also identification of relevant dimensions, goals, and KPIs. For this reason we set specific dimensions in Industry 4.0 rather than using the typical dimensions of the Balance Scorecard.

3.3.1 Dimensions, KPIs, and Goals Identification

We analyzed the literature on Smart Production, in particular we searched the term "Smart Factory" in Google Scholar. By looking the definitions of a Smart Factory, we are able to understand the desired Smart Production process in the context of Industry 4.0. The collected definitions were dissected and classified into three dimensions: Humans, Machines, and Systems. We describe each dimension as follows:

- The **Humans** dimension is related with activities performed by workers of a Smart Factory or activities performed by customers or contractors of a Smart Factory.

Table 3.1 KPIs in Smart Factory

Goal	Dimension	KPI	Values
Context awareness [25]	Machine	Amount of sensors	Integer
Context awareness [25]	Machine	Sensor availability	0 or 1
Virtual representation [26]	Machine	Amount of virtual representations	Integer
Virtual representation [26]	Human	Digital design	0 or 1
Demand orientation [26]	System	Personalization availability	0 or 1

- The **Machines** dimension is related with the machines of a Smart Production process.
- The **Systems** dimension is related with the software components of a Smart Production process.

The outcome of the dissection and classification process resulted in a list of key attributes (Table 3.1). Those key attributes were converted into key performance indicators (KPIs) by adding a measuring attribute such as "number of" or "availability." The "number of" measuring attribute receives an integer value and the availability measuring attribute receives a binary value of zero or one. The value is zero if this KPI is not implemented in the production process and the value of the KPI is one when it is implemented. These binary values are helpful for transitioning from a traditional manufacturing process towards a Smart Production process.

The converted KPIs are to be interpreted as follows:

- The **sensor availability** reflects the overall potential of the production process digitization and automation. The availability of sensors is required for the implementation of cyber-physical production systems.
- The **amount of sensors** represents the number of sensors that serve as potential digital interfaces for a Smart Production process. The higher the number, the better is the automation potential due to an increased fidelity of the physical production process with a digital one.
- The **amount of virtual representations** tells us how many different variants are supported by the process.
- The **digital design** tells us whether the humans can influence production planning process or some part of it.
- The **personalization availability** tells us whether the process or some part of it is compatible with customization.

3.3.2   Selection of the Use Case

In order to test the identified goals, KPIs, and dimensions presented in Table 3.1 we have chosen a manufacturing process from the industry previously introduced by [27] presented in Fig. 3.1 that represents a very generic use case for variant produc-

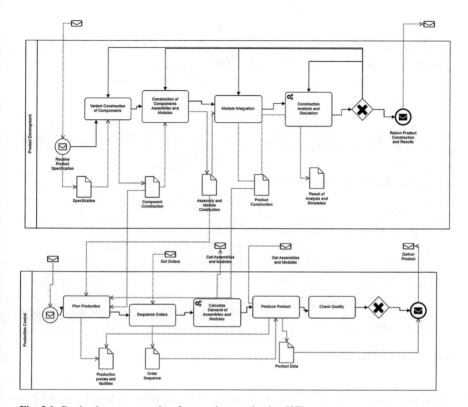

Fig. 3.1 Production use case taken from variant production [27]

tion. This use case is appropriate since all identified KPIs can be demonstratively assigned to the specific process tasks.

3.4 Results

In following subsection we present two thinkable scenarios for process tracking in Smart Production. The first one relies on standalone application as previously introduced in [28]. The second one involves an IoT infrastructure and enables live tracking of KPIs.

3.4.1 Solution as Standalone Application

Once the process is modeled and available in BPMN we can load it into management tool. First step requires initial definitions of the actors and goals, dimensions, and

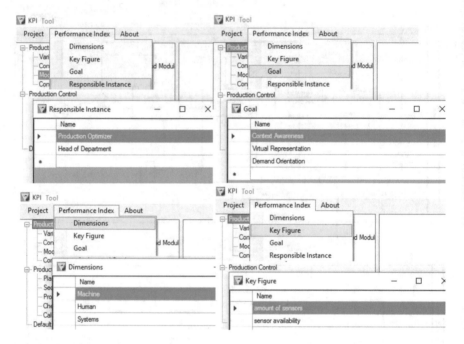

Fig. 3.2 Definition of KPIs

KPIs we defined in Table 3.1. Figure 3.2 shows the definition assigned according to the KPIs.

Tracking the KPIs

Figure 3.3 shows this workflow and the direct assignment of tracking entries to the tasks in the process. By clicking in the tree view or by direct selection in the process model we can swiftly get an overview about already defined KPIs for a specific task and edit or delete them. In this way, it is possible to continuously update the status of KPIs using the KPI Tool along the production process.

To get an overview over all defined KPIs and their values, KPI Tool supports listing of them within the application as BSC table (see Fig. 3.4) but also as an export option (e.g., as table of comma separated values).

The exported values can be used, for instance, for visual overview and tracking in other software tools as depicted in example in Fig. 3.5. In this case we can use parallel coordinates to visually explore the assigned values to the KPIs and to filter them on demand. In this way we gain a fast overview on KPIs and we are able to drill-down the specific views via the visual interface.

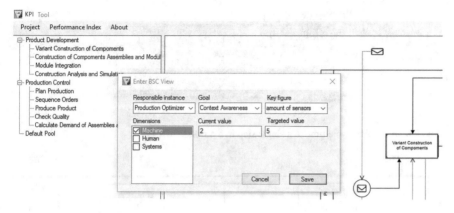

Fig. 3.3 Assignment of KPIs

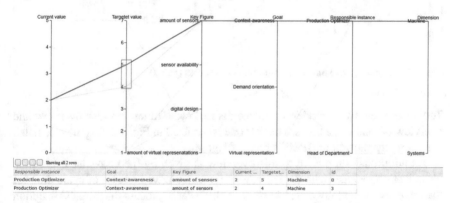

Fig. 3.4 Overview with all assigned KPIs in the process

Fig. 3.5 Overview with parallel coordinates visualization

Tracking the Cycle Time

Additionally to the regular KPI tracking the KPI Tool contains an option to track the cycle time per process and per step described as expression of time value and probability of the (re-)occurrence for the given process step. In our case, we can also say production step. The assignment process of cycle time for a single step is shown in Figs. 3.6 and 3.7.

As we can see in Fig. 3.7 the cycle time value can be expressed also as calculation derived from the single task and the structure of the BPMN process. It is represented as calculation expression and it can be easily assigned by the entering mask. The KPI

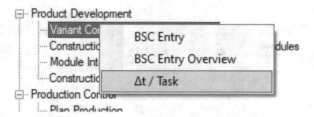

Fig. 3.6 Assigning cycle time to a single step in a process (part 1)

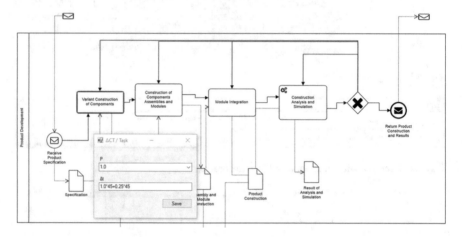

Fig. 3.7 Assigning cycle time to a single step in a process (part 2)

Tool also offers two overviews over the defined cycle times: per task overview and overview of total cycle times. The first one is depicted in Fig. 3.8. It is the calculated view directly showed in BPMN process diagram.

An additional alternative to this overview is given by the task-wise cycle time visualization together with depiction with the cumulative increase of cycle time for the traversed path along a certain route in the BPMN process diagram. This option is depicted in the Fig. 3.9.

As we can see on presented examples the cycle time tracking option allows trying out different setups in combination with time assignments. This is especially useful when, for instance, a production optimizer plans the changes in production line (process form changes) and wants to simulate manually the impact of process changes on the overall production time.

3.4.2 Solution as Part of IoT Infrastructure

Figure 3.10 shows the concept of an extended version of tracking scenario which involves also an IoT infrastructure. The basic idea is to change the parameters using

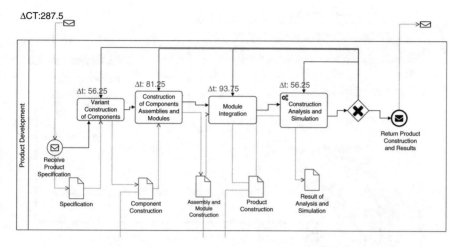

Fig. 3.8 In diagram overview over single and total cycle times

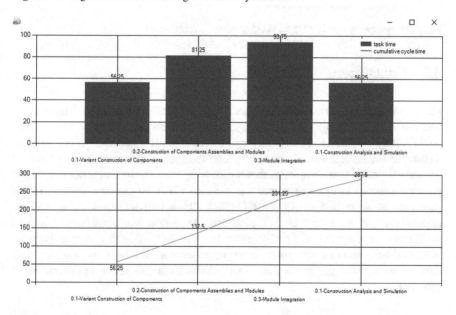

Fig. 3.9 Task-wise and cumulative overview over cycle times in a process

the signals from sensors and directly from production line. In presented scenario in Fig. 3.10 we have sensors in production line involved and connected via wireless Internet to an IoT infrastructure. The infrastructure consists from data storage for recording sensor statuses and from web service interface exposed to KPI Tool. In this ways the number of active sensors in variant production is updated and set on-line.

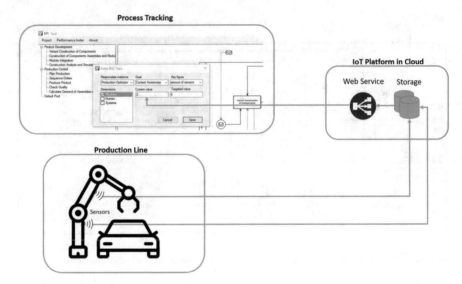

Fig. 3.10 Assigning and tracking KPIs using the IoT infrastructure

3.5 Conclusion

Our novel approach combines KPIs tracking with Smart Production. This tool let users to track the change of KPIs from an as-is scenario to a desired to-be scenario. Additionally we can assign and track cycle times for each step and overall process. Those are the crucial contributions of presented tool. This comparison could help traditional production companies that are shifting towards Smart Production to test possible process setups and define the indicators that should be considered. The presented solution in form of the KPI Tool allows two scenarios: as standalone application and as a part of IoT platform. As future work, we would like to implement an historical KPI performance analysis to be able to compare the prior values of KPIs and to detect some trends or patterns as well as to use the historical data as base input for predictions that allows us a pro-active optimization of processes.

References

1. S. Waltzinger, P. Ohlhausen, D. Spath, The industrial internet. Business models as challenges for innovations, in *23rd International Conference on Production Research, ICPR 2015*, Manila (2015), https://core.ac.uk/download/pdf/45359614.pdf
2. C. Zott, R. Amit, L. Massa, The business model: Recent developments and future research. J. Manage. **37**(4), 1019–1042 (2011). https://doi.org/10.1177/0149206311406265
3. S. Zor, D. Schumm, F. Leymann, A proposal of BPMN extensions for the manufacturing domain, in *Proceedings of the 44th CIRP International Conference on Manufacturing Systems* (2011)

4. T. Allweyer, *BPMN 2.0: Introduction to the Standard for Business Process Modeling* (Books on Demand, Norderstedt, 2009)
5. I. Graja, S. Kallel, N. Guermouche, A.H. Kacem, BPMN4CPS: A BPMN extension for modeling cyber-physical systems, in *2016 IEEE 25th International Conference on Enabling Technologies: Infrastructure for Collaborative Enterprises (WETICE)*, pp. 152–157 (2016). https://doi.org/10.1109/WETICE.2016.41
6. R. Petrasch, R. Hentschke, Process modeling for industry 4.0 applications: towards an industry 4.0 process modeling language and method, in *2016 13th International Joint Conference on Computer Science and Software Engineering (JCSSE)*, pp. 1–5 (2016). https://doi.org/10.1109/JCSSE.2016.7748885
7. R.S. Kaplan, D.P. Norton, *The Balanced Scorecard: Translating Strategy Into Action* (Harvard Business School Press, Boston, MA, 1996)
8. F. Al-Turjman, M.Z. Hasan, H. Al-Rizzo, Task scheduling in cloud-based survivability applications using swarm optimization in IoT. Trans. Emerg. Telecommun. Technol. **30**(8), e3539 (2019). https://doi.org/10.1002/ett.3539
9. F. Al-Turjman, A. Malekloo, Smart parking in IoT-enabled cities: a survey. Sustain. Cities Soc. **49**, 101, 608 (2019). https://doi.org/10.1016/j.scs.2019.101608
10. F. Al-Turjman, Intelligence and security in big 5g-oriented IoNT: an overview. Futur. Gener. Comput. Syst. **102**, 357–368 (2020). https://doi.org/10.1016/j.future.2019.08.009
11. F. Al-Turjman, M. Abujubbeh, IoT-enabled smart grid via SM: an overview. Futur. Gener. Comput. Syst. **96**, 579–590 (2019). https://doi.org/10.1016/j.future.2019.02.012
12. F. Al-Turjman, H. Zahmatkesh, L. Mostarda, Quantifying uncertainty in internet of medical things and big-data services using intelligence and deep learning. IEEE Access **7**, 115, 749–115, 759 (2019). https://doi.org/10.1109/ACCESS.2019.2931637
13. E. Lüftenegger, S. Softic, Service-dominant business model financial validation: Cost-benefit analysis with business processes and service-dominant business models, in *Proceedings of 30th Central European Conference on Information and Intelligent Systems (CECIIS 2019)*, ed. by V. Strahonja, V. Kirinic. University of Zagreb, Faculty of Organization and Informatics, Varazdin (2019)
14. M. Chen, S. Mao, Y. Liu, Big data: a survey. Mob. Netw. Appl. **19**(2), 171–209 (2014). https://doi.org/10.1007/s11036-013-0489-0
15. S. Softic, M. Zoier, A. Stocker, Big data. mit sprechenden daten zu optimierten geschäftsprozessen. Virtual Veh. Mag. **1**(20), 16–17 (2014)
16. E.W.T. Ngai, A. Gunasekaran, S.F. Wamba, S, Akter, R. Dubey, Big data analytics in electronic markets. Electron. Mark. **27**(3), 243–245 (2017). https://doi.org/10.1007/s12525-017-0261-6
17. W. Bauer, S. Schlund, D. Marrenbach, O. Ganschar, Industrie 4.0 – Volkswirtschaftliches Potenzial für Deutschland. Studie, Bundesverband Informationswirtschaft, Telekommunikation und neue Medien e.V. (Bitkom) mit dem Fraunhofer-Institut für Arbeitswirtschaft und Organisation (IAO, Stuttgart), Berlin (2014). https://www.bitkom.org/Bitkom/Publikationen/Industrie-40-Volkswirtschaftliches-Potenzial-fuer-Deutschland.html
18. W. Becker, P. Ulrich, T. Botzkowski, *Industrie 4.0 im Mittelstand - Best Practices und Implikationen für KMU*, 1st edn. (Springer, Berlin/Heidelberg/New York, 2017)
19. A. Borgmeier, A. Grohmann, S.F. Gross, *Smart Services und Internet der Dinge: Geschäftsmodelle, Umsetzung und Best Practices* - Industrie 4.0, Internet of Things (IoT), Machine-to-Machine, Big Data, Augmented Reality Technologie (Carl Hanser Verlag GmbH Co KG, 2017)
20. T. Kaufmann, *Geschäftsmodelle in Industrie 4.0 und dem Internet der Dinge - Der Weg vom Anspruch in die Wirklichkeit*, 1st edn. (Springer, Berlin/Heidelberg/New York, 2015)
21. E. Lüftenegger, *Service-Dominant Business Design*. Eindhoven University of Technology (2014), https://doi.org/10.6100/IR774591
22. W. Brenner, T. Hess, W. Brenner, T. Hess, *Wirtschaftsinformatik in Wissenschaft und Praxis - Festschrift für Hubert Österle*, 1st edn. (Springer, Berlin/Heidelberg/New York, 2014)
23. R. Wieringa, *Design Science Methodology for Information Systems and Software Engineering* (Springer, Berlin, 2014). https://doi.org/10.1007/978-3-662-43839-8

24. A.R. Hevner, S.T. March, J. Park, S. Ram, Design science in information systems research. MIS Q **28**(1), 75–105 (2004). http://dl.acm.org/citation.cfm?id=2017212.2017217
25. D. Lucke, C. Constantinescu, E. Westkämper, Smart factory - a step towards the next generation of manufacturing, in *Manufacturing Systems and Technologies for the New Frontier*, ed. by M. Mitsuishi, K. Ueda, F. Kimura (Springer, London, 2008), pp. 115–118
26. J. Lee, Smart factory systems. Informatik-Spektrum **38**(3), 230–235 (2015). https://doi.org/10.1007/s00287-015-0891-z
27. D. Roller, E. Engesser, BPMN process design for complex product development and production, in *Informatik 2014*, ed. by E. Plödereder, L. Grunske, E. Schneider, D. Ull (Gesellschaft für Informatik e.V., Bonn, 2014), pp. 1979–1984
28. E. Lüftenegger, S. Softic, S. Hatzl, E. Pergler, A management tool for business process performance tracking in smart production, in *Mensch und Computer 2018 - Workshopband*, ed. by R. Dachselt, G. Weber (Gesellschaft für Informatik e.V., Bonn, 2018)

Chapter 4
Trust-Based Chaos Access Control Framework by Neural Network for Cloud Computing Environment

J. V. Bibal Benifa and G. Venifa Mini

4.1 Introduction

Cloud computing is an emerging technology that offers unmatched computing experience to users with unconstrained computing resources in a cost effective manner [1]. The cloud infrastructure offers flexible and on-demand based computing resource allocation to end users [2]. It enables the user to reduce hardware infrastructure and maintenance costs intended for computing. Cloud computing offers numerous security features for the end users that include data security and privacy which are considered as the two major parallel concerns [3, 4]. Apart from these concerns, trust is an important attribute in the cloud environment and it is applied at several stages of the cloud architecture. Here, trust between data users & the data content as well as the trust between the data proprietor and cloud server are the two important stages. It is often related to the reliability of the CSP who is proficient to offer the agreed service capability, security and privacy [5, 6]. In addition, trust of data user is related to the nature of data usage such as usage of data solely for the intended purpose. Hence, trust is not a single dimensional attribute rather it is achieved by the combination of several features ranging from data security to privacy. As the matter of fact, a problem occurred at any one of the stages in the cloud architecture may directly influence the trust factor [7–9]. Since, cloud trust is directly proportional to data security and privacy; inadequate security features directly affect the trust factor and quality of service [10, 11]. Lack

J. V. B. Benifa (✉)
Department of Computer Science and Engineering, Indian Institute of Information Technology, Kottayam, India

G. V. Mini
Department of Computer Science and Engineering, Noorul Islam Centre for Higher Education, Kumaracoil, India

© Springer Nature Switzerland AG 2020
F. Al-Turjman (ed.), *Trends in Cloud-based IoT*, EAI/Springer Innovations
in Communication and Computing, https://doi.org/10.1007/978-3-030-40037-8_4

of control over data content and transparency during the processing of data by the CSP diminishes the trust factor [12].

To authorize user access, CSP assigns a trust level based on the previous behavioral history of the corresponding user. CSP allows the authorized user with a high level freedom to data access at various unanticipated situations. On the other hand, the unauthorized user could be a malicious user, hacker or potential competitor who involves vulnerable policy violations. The impact of attack causes serious damage in terms of cost or abolishes the life of CSP. Therefore, timely detection and prevention of malicious and unauthorized user access is an important aspect in cloud computing environment. In the present chapter, a comprehensive trust evaluation module is proposed with necessary architecture and implementation scheme for efficient trust monitoring and control in cloud computing environment.

4.2   Related Work

Trust is accepted as a strong factor in the growth of usage of cloud computing-based services. It is essential for any such factor should be quantifiable in one or more means, and numerous studies have been conducted in the recent years to evaluate the trust factor [13–15]. In particular, a research work done by Tang et al. [16] to evaluate and monitor the trust in the cloud environment employs a trust monitoring middleware. The proposed middleware was interfaced along with the common interfaces of cloud computing such as client-server, owner-server, server-server and server-processing CSP. The evaluation is initialized with trustworthiness and QoS monitor which is evaluated with the help of a synthesized web services dataset. The results are quantified in terms of cost, time and speed of the particular cloud service. In addition, user ratings and feedbacks also subjectively measured to evaluate the trustworthiness of the service being offered by CSP. Few main factors considered by Tang et al. for measuring the trust are QoS conformity, percentage of untrustworthy users, throughput of Service Level Agreements (SLA), and reliability of cloud services. Wang et al. proposed a trust and privacy aware cloud computing evaluation model. The authors suggest that the trust factor is a dynamic quantity with respect to time; hence they introduced a time specific trust model that changes with respect to time [17]. A simulation is performed for the proposed trust and privacy aware cloud monitoring model with the help of MATLAB. Trust is directly measured between the satisfaction levels of the end user and plotted against the total number of transactions. Further, the number of transactions vs. the success rate of transactions also presented in the analysis. The work also simulates a number of malicious cloud services and their success rate with optimum accuracy.

The dynamic cloud resources require more serious evaluation structures so that it can be adapted to the changing user and service provider's scenario. A feedback-based trusted service provider selection module was proposed by Varalakshmi et al. [18]. The proposed system is competent for removing feedbacks by the malicious users and attains accurate trust-based service provider identification. The dynamic nature of trust factor is well understood by the authors as they have included a dynamic trust capability module. In the proposed system, a universal trust management module is enacted that utilizes the underlying platform information during processing and thereby updates the trust monitoring module. The universal advisory module shall be interconnected with a number of CSP, whose service level performance will be continuously monitored by the universal module. A new concept trustworthiness agreement is also included in this work which is a mutual consent on agreed terms regarding the trust among CSP or clients. A user who intends to use the cloud service may be unaware of the trustworthiness of the CSP and initializes a request with the universal advisory, which in turn possesses previous user's feedback. It also monitors the performance and quality of CSP who accomplishes the user's request with trust-related information. The overall system provides trustworthy service provider selection platform and the internal trust features such as data confidentiality, data security, etc. which are not discussed in the aforementioned work.

A compliance-based CSP trust assessing method was proposed by Sarbjeet et al. [19]. Trust was measured in terms of QoS between the cloud client and server, cloud broker and server, and among the servers. Some of the elements of consideration for the measurement of trust are availability of service, reliability, response time, flexibility of cloud services, usability, confidentiality, scalability, and transparency. A QoS matrix is formed with the elements of trust measurement and cloud auditor module that monitors every trust parameter and ensures trust scoring for the particular module in the cloud system. The performance of the module elements for trustworthiness will be evaluated by the cloud auditor and the results shall be given to intended users on request basis. However, in this work trust is solely considered as a matter of service quality rather than data security and confidentiality.

4.3 Influence of Neural Network in Trust

4.3.1 Neural Network Classifiers

Artificial neural networks (ANN) is derived concepts from their biological counterparts such as the neurons in the brain [20]. ANN concept enables computing systems to observe, learn, and take decisions from observational data [21]. It provides a number of solutions to technological problems arising in engineering and day-to-day life as well [22]. Further, it is proficient to solve technological issues in image recognition, speech recognition, pattern identification, classification, grouping, and natural language processing [23]. To deeply understand the concept of NN, it is essential to know about the basic components of the technique. Among the basic components, perceptron is a type of artificial neuron developed in early stages of development of NN concepts. A perceptron takes a number of binary inputs, x_1, x_2, and produces a single binary output.

Fig. 4.1 A simple perceptron

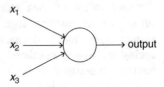

In Fig. 4.1, a perceptron takes three inputs, namely x_1, x_2, x_3, for a single output. Normally, the inputs can be of any numbers; however, it produces single output only. Perceptron produces output based on rules, where it applies rules among the inputs and identifies their inter-relation and produces a single result. Rules are technically denominated as weights in NN concepts. The weights are denoted as w_1, w_2, w_3. The weights are represented by a real number such that they represent the priority and essentiality of the inputs with respect to the output. A neuron's output 0 or 1 is calculated, if the weighted sum $\sum_j w_j x_j$ is less than or greater than the threshold value. The threshold value is also a real number which is one of the factors of the neurons. Mathematically, neurons can be represented as follows in Eq. (4.1),

$$\text{output} = \begin{cases} 0 \text{ if } \sum w_j x_j \leq \text{threshold} \\ 1 \text{ if } \sum w_j x_j \geq \text{threshold} \end{cases} \tag{4.1}$$

The above mathematical representation provides a deep understanding about the working of the neuron. The neuron produced a result 0, if the weighted sum is less than or equal to the threshold value. The neuron produces result 1, if the weighted sum is greater than the threshold value. The weights in ANN w_1, w_2, w_3 ... determine how the input alters the output [24]. Weight can be adjusted or changed so that a particular input gets more priority than the other inputs or vice versa. If the weight is varied, then the output of the network also changes in a NN [25].

A NN can also be more complex as presented in Fig. 4.2. This representation consists of a multistage network. At the first stage, a number of inputs are processed with three weights and outputs are produced. In the next stage, four weights are applied to the processed inputs which finally produce a single output. In this network, the first column of perceptron is called first layer of perceptron. This first layer produces simple outcomes by weighing the input data. The next column of perceptron adjacent to the first level is called second layer of perceptron. They decide the outcomes at a complex level than the first layer of perceptron. The third layer of perceptron is meant for more complex decisions than the first two layers. Hence, a number of multilayer network of perceptron can work together to produce a refined decision-making.

Fig. 4.2 Complexity of
neural networks

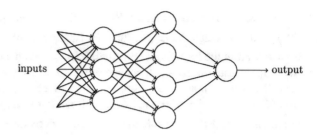

4.3.2   Back-Propagation Algorithm

NN learns their weights using gradient descent algorithms and they calculate the gradient of a cost function with any specific algorithm [26]. One among such algorithms for computing gradient is back-propagation algorithm (BPA). BPA consists of two stages, namely, feed-forward stage and back-propagation stage. As given in algorithm, examples are represented with input vector x and output vector y. In the feed-forward stage, each layer values are calculated from weight $W_{j,i}$, activation function g, and values from the previous layer. It turns the error from output to the hidden layer. Back-propagation-based algorithm produces faster outcomes by the way of solving problems than any other learning algorithms developed previously.

Algorithm 1 Back-Propagation Neural Network Learning Algorithm
function BACK-PROP (*examples, network*) **returns** a neural network
    inputs: *examples*, a set of examples, with input vector x and output vector y
           network, network with L layers, weights $W_{j,i}$, activation function g
    repeat
    for each e **in** *examples***do**
        for each node j in the input layer **do** $a_j \leftarrow x_j[e]$
        // **Feed Forward Stage**
        for $l = 2$ **to** M**do**
            $in_i \leftarrow \sum_j w_{j,i} a_j$
            $a_i \leftarrow g(in_i)$
        //**Back-Propagation Stage**
        for each node i in the output layer **do**
            $\Delta_i \leftarrow g(in_i) \times (y_i[e] - a_i)$
        For $l = M - 1$ **to 1 do**
            for each node j in layer l **do**
                $\Delta_i \leftarrow g(in_i) \sum_i w_{j,i} \Delta_i$
                for each node i in layer $l + 1$ **do**
                $w_{j,i} \leftarrow w_{j,i} + \alpha \times a_j \times \Delta_i$
    Until stopping criteria
    return *network*

The very basic core behind BPA is the formula for the partial derivative $\partial C / \partial w$ that belongs to the cost function C, for any weight w given in a particular network. From the formula, it is well understood that the cost changes rapidly with the weight variations. A small change in weight leads to a substantial change in the cost. The BPA computes cost from weights and it also provides a description of the overall characteristics of the network with respect to its weights. Hence, the main intention of the BPA is to calculate the partial derivatives $\partial C / \partial w$ and $\partial C / \partial b$ of the cost function C for any weight w in the network. The quadratic expression of the cost function is expressed as follows in Eq. (4.2),

$$c = \frac{1}{2^n} \sum_x \| y(x) - a^L(x) \|^2 \tag{4.2}$$

In the above expression, n is the total number of training examples; sum of individual training sets, x; $y = y(x)$ belongs to its respective output L; L is the total number of layers in the network; $a^L = a^L(x)$ is the vector activations output from the network for the input x.

4.4 Trust-Based Chaos Access Control (TBCAC)

The cloud Trust Manager (TM) is an integral part of the proposed system for enhancing security, privacy, and trust. The key purpose of the TM is to monitor, analyze, and enforce the trust requirements on data owner's data publication. Primarily, the trust monitoring is done based on the locality and how the data is being utilized by the cloud users. Subsequently, the TM receives a copy of user's profile including their role and permission privileges. Along with the profile, the TM would be aware of the data requesting pattern, online time duration, volume of data transaction, and sensitivity of such data. Then, the TM classifies the user with the help of NN-based classifier according to their trust level. A range of trust levels are predefined in the architecture which can be altered according to the data owner's need. Once the user is classified, he will be restricted from accessing data if he is categorized as a low trustable user. On the other hand, a user continues to access his allotted roles and privileges, if he maintains his trust level.

The important feature of the TM is the dynamic nature and controlling capability. It runs in a real-time environment and accesses trust-related data on real time with respect to each and every cloud user's transaction on the cloud server. The cloud manager is capable of dynamically analyzing the data requesting pattern from the user with the help of NN-based classifiers. The NN-based classifiers have internal forward error correction control which means that it has the potential to correct by its own over each and every classification. Finally, the classification results in the categorizing of user to trust levels. There are three types of trust levels which are predefined in the manager known as low trustable, moderate trustable, and highly trustable. A low trustable user level is subjected to restricted data access irrespective of its roles and permissions. A moderately trustable user will continue to access the allotted privileges; however it is subjected to the consideration by data owner. A highly trustable user will be continued to access the data without any restriction being imposed over its roles and permissions. The system architecture and its corresponding embodiments are presented in Fig. 4.3.

The TM is an integrated part of the proposed cloud computing system as it runs from the CSP server. The TM consists of a data acquisition module that receives data required to identify the trust level of the data user. The data consists of user access roles, privileges, and permissions. Apart from the basic data, the TM also receives the user access behavioral data from time to time. The user access behavioral data consists of user access pattern, access privilege violation attempts, session duration on the particular data set, frequency of data access, and volume of data access. TM collects the user behavioral data for every fresh session as the cloud user accesses the server. Thus, the acquired behavioral data will be pipelined for classification with the help of a NN-based classifier. Data pipelining is an essential component as it ensures minimal handling of resources used by the TM. Data pipelining also ensures the total number of NN-based classifiers which are in place to cater the requirement of timely classification of large number of users per session.

The primary module of TM is the ANN-based classifier and it consists of a data feed module. This module acts as a buffer for the storage of data to be used by the classifier. The feed will send the data to the classifier which works based on BPA. Subsequently, the errors obtained during classification will be back propagated and used for enhanced classification during the next iteration. Post classification process, the classification outcomes will be stored in the cloud server for future reference. Logging enables the TM to keep track of trust history of a user. Based on the post logging of trust outcomes, low trustable users will be restricted from accessing data from the server.

Moderately trustable users will be allowed to access the data, simultaneously a notification will be sent to the data owner for reconsideration of roles and privileges of such moderately trustable users if required. The data owner may change or revoke role privileges of moderate trustable users or may allow the user to continue with the current role based on his sole discretion. Hence, the low trustable users will be restricted immediately from accessing data from the cloud server. It is important to note that, a low trustable user will not be allowed to access data from the server unless the data owner grants exclusive access permissions. Such dynamic identification and isolation of low trustable users improve the overall system confidence and reliability. It prevents unauthorized and non-trustable users from accessing cloud data.

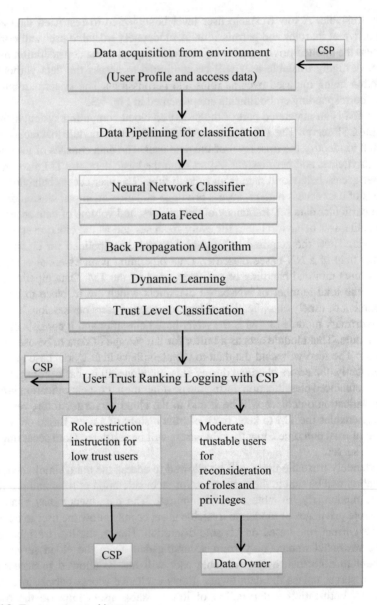

Fig. 4.3 Trust manager architecture

4.4.1 *Trust Manager Work Flow Structure*

In Fig. 4.4, the overall work flow of the TM is detailed, wherein the TM receives user profile and transaction-related data for each and every session of the cloud user from the CSP server. The TM has buffer storage provisions for storing input and

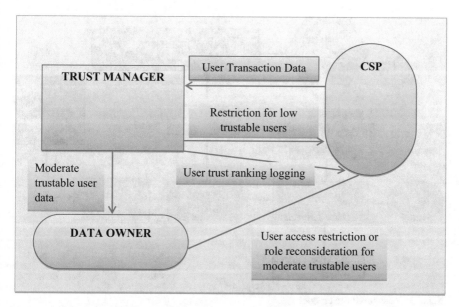

Fig. 4.4 Work flow diagram of trust manager in cloud computing

processing data for the specific time period. TM executes the classification with the help of NN-based classifiers and categorizes the user data into their corresponding trust level. Three levels of trusts are defined for the proposed system as stated in Sect. 4.4. Once the classification process is completed, the TM logs the outcome in the server prior to sending restriction instruction to the CSP for forbidding access to the low trust users. Further, the moderate trustable users are notified to the cloud data owner for necessary action. The moderate trustable users are reviewed by the data owner and their corresponding access privileges may be revoked and reconsidered if they violate the trust.

The TM has a predefined threshold value for each and every trust level ranking which may be changed by the data owner periodically.

4.4.2 Role-Based Trust Access Control

The proposed model for user requests to access various resources from cloud environment is depicted in Fig. 4.5. Here, it verifies the user behavior stored in access log and calculates the trust value of the user. The trust calculation is based on behavior monitor data, and according to the information, a user is classified as trusted or an untrusted user. Subsequently, the dynamic access control module assigns the user with roles depending on the trust value. The user obtains authority based on roles and the roles perform operations on specified objects. At the next time, this module recalculates trust value and assign new role to the user. According

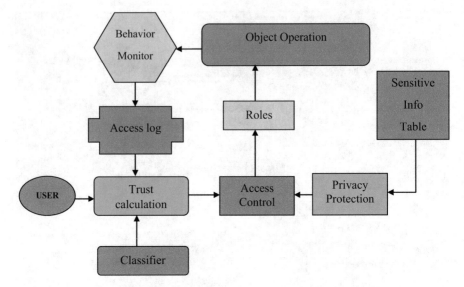

Fig. 4.5 Role-based trust model

to the trust value and past behavior history, the role and priority are assigned to the user. Based on this assessment, CSP can provide flexible access control.

4.4.3 User Behavior Trust Calculation

Trust is not calculated solely on the current session parameters, instead it is consolidated from the previously calculated trust values as shown in Fig. 4.6. The weighting method is applied while calculating the consolidated trust; weight factor is applied between past and current trust values such that the overall user behavior throughout his lifetime with the cloud is taken into account. If total behavior trust value is higher than the defined threshold, the user is trustworthy enough to perform an action on a certain object. User behavior trust is calculated by analyzing current trust value and past trust value. Trust is calculated based on weights given to four important parameters, namely session (α_1), volume of data access (α_2), violation attempts (α_3), and number of accesses (α_4).

4.4.4 Current Trust (Currtrust)

Current Trust is calculated as the geometric mean of the above parameters,

$$\left(\pi_{i=1}^{n} \alpha_i\right)^{1/n} = \sqrt[n]{\alpha_1 * \alpha_2 * \ldots \cdots * \alpha_n} \tag{4.3}$$

Fig. 4.6 Trust consolidation

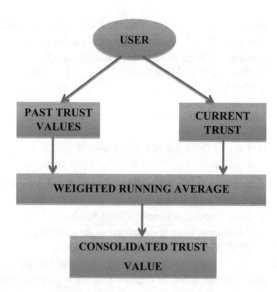

where α is the individual parameter and n is the number of parameters. In the proposed system, the number of parameters involved is 4. The geometric mean compares different parameters and finds a single merit for these parameters.

Based on the above generalized equation trust calculation for the proposed system is given as follows:

$$\text{Trust} = \sqrt[4]{\alpha_1 \times \alpha_2 \times \ldots \alpha_4} \qquad (4.4)$$

4.4.5 Past Trust Value (Pasttrust)

The past trust value acts as a supporting factor for trust calculation when the tries made to access the resources in the next attempt. Past trust value is also considered as the consolidated trust value from previous transactions.

4.4.6 Consolidated Trust Value (Construst)

The consolidated trust values verify that the user is trusted or untrusted user. The consolidated trust value is a function of current and past trust values.

$$\text{construst}(i) = \alpha \times \text{construst}(i) + \beta \times \text{pasttrust}(i) \qquad (4.5)$$

where

construst(i) is the consolidated trust of ith transaction
currtrust(i) is the current trust of ith transaction
pasttrust(i) is the past trust of ith transaction
α Constant weighting factor $0 \le \alpha \le 1$ to the current trust
β Constant weighting factor $1 - \alpha$ to the past trust

4.4.7 Trust-Level Threshold

According to the evaluation of consolidated trust value, the levels of trust factor is expressed as follows:

Low Trust if construst < thresholdtrust.

Moderate Trust if construst = thresholdtrust.

High Trust if construst ≥ thresholdtrust.

A simple method is used to evaluate the trustworthiness of user. If the consolidated trust value is greater than or equal to threshold trust, which is 2.5 based on arithmetic mean of past and current trust, then he/she is a trusted user. If the consolidated trust value is lesser than the threshold trust, then he/she is an untrusted user. After the evaluation of consolidated trust value, that value is forwarded to the CSP for final decision regarding access control.

4.4.8 Break the Glass Approach

The TM regulates and monitors the users for trust enforcement as presented in Fig. 4.7. However, it plays an important role for implementing greater flexibility during an emergency scenario. If a high trust user wants to access the data belonging to data owner in an emergency scenario, then the trust authenticates the user with third party authorities or with law enforcement agency based trusted authorities. Once authenticated, the particular user is allowed to access the data with a break the glass provision on the emergency situation.

On the other hand, the data owner can also define emergency access policies by registering it as a role. In this case, the data owner will have a separate key stored in the key management module which is normally attributed to the physiological parameters of the data owner such as ECG pattern or thump impression. If a health care provider wants to access the data owner's private health recorder in an emergency situation, he may use the data owner's physiological data to authenticate with the key manager and access the data with break the glass provision. This provision adds greater flexibility to the access control in an emergency situation.

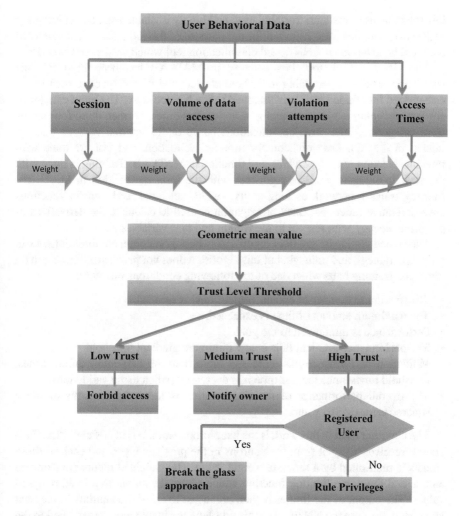

Fig. 4.7 Trust-level threshold structure

4.5   Experimental Evaluation

The simulation of the proposed system is carried out with MATLAB software on an Intel core $i5$ processor based computer system and the Amazon cloud services dataset information is user for trust calculation [27, 28]. The primary data involved in the trust factor determination are session of a particular user with the cloud server in milliseconds, volume of data accessed by the user during his lifetime in Kilobytes, number of access privilege violations attempted by the particular user in lifetime, and number of times of accesses by the user in lifetime. According to these primary data, trust level of the users will be categorized into three levels, namely, low trustable user, medium trustable user, and highly trustable user. Data pertaining to 200 user samples were recorded for the purpose of this simulation. The recorded data will be given to an ANN-based classification tool which works based on BPA.

The NN Simulink tool box running in MATLAB has been used for this simulation. The data pertaining to 200 cloud users and their respective trust levels categorized into three levels was assigned to the NN tool box. The input categories are session, volume, violation attempts, and access times. Further, these four inputs were categorized into three classifications based on 15 internal weights that are hidden in the NN. Data division for training, validation, and test are done with random selection with the *divider and* function in MATLAB. Training is done with the scaled conjugate gradient algorithm with the *trainscg* function in MATLAB. *trainscg* trains a network as long as its weight, net input, and transfer functions have derivative functions. Back propagation is used to calculate the derivatives of performance *perf* with respect to the weight and *bias* variables X.

The scaled conjugate gradient algorithm is based on conjugate directions, as in traincgp, traincgf, and traincgb, but this algorithm does not perform a line search on iteration. Training halts when one of the following conditions met with:

- The maximum number of epochs (repetitions) is reached.
- The maximum amount of time is exceeded.
- Performance is minimized to the goal.
- The performance gradient falls below minimum gradient threshold.
- Validation performance has increased more than the maximum failure times, threshold times since the last time it is decreased (when using validation).
- The overall performance of the classification is measured in terms of mean squared error (MSE) value.

The classifier used in this work is the back-propagation-based NN classifier. Each neuron receives a signal from the neurons in the previous layer, and each of those signals is multiplied by a separate weight value. The weighted inputs are summed and passed through a limiting function which scales the output to a fixed range of values. The output of the limiter is then broadcast to all of the neurons in the next layer. Hence, to use the NN to solve this problem, the input values are applied to the first layer and it permits the signals to propagate through the network, and reads the output values. Simulation is initiated with the inputs of the first layer, and signals propagate through the central (hidden) layer(s) to the output layer. Each link among the neurons has a distinctive weighting value and inputs from one or more preceding neurons are individually weighted, then summed.

The end result is nonlinearly scaled between 0 and +1, and the output value is passed on to the neurons in the successive layer. Since the real distinctiveness or "intelligence" of the network presents in the values of the weights between neurons, a scheme of adjusting the weights is required to solve a specific problem. For this type of network, the most common learning algorithm BPA is used. A back-propagation network learns by paradigm that provides a learning set consisting of some input examples and the known-correct output for each case. Therefore, these input-output examples are exploited to show the network that what type of behavior is anticipated, and the BPA permits the network to adapt accordingly.

The BPA learning method works through few iterative steps: one of the example cases will be applied to the network; it generates some output corresponding to the existing state of its synaptic weights (initially, the output will be random). This output is evaluated to the known-good output, and a mean-squared error signal is computed. The error value is then propagated backwards through the network, and slight modifications are made to the weights in each layer. The weight variations are calculated to minimize the error signal for the specific case according to the queries. The entire process is repeated for each of the example cases, and again back to the first case once more, and so on. This is a cyclic process until the overall error value drops under the predetermined threshold. At this point, one can state that the network has learned the problem "well enough" and it will never learn the ideal function, but rather it approaches the ideal function asymptotically.

4.5.1   Results and Analysis

The results obtained from the MATLAB NN toolbox are discussed in this section. A histogram is a graphical depiction of the distribution of numerical data. It is an approximation of the probability distribution of a continuous variable which can be represented as a bar graph. To build a histogram, the initial step is to "bin" the range of values, i.e., divide the full range of values into a series of intervals and then count up how many values fall into each interval. The bins are frequently specified as consecutive, nonoverlapping intervals of a variable. The bins should be adjacent and often assumed with equal size. The error occurring interval is also plotted in a histogram specified in Fig. 4.8.

In the present analysis, at the interval -0.04955 where around 380 instances occurred and the probability of error is zero. This is the optimum interval of the classification using NN classifier. The training state statistics show that the NN classifier achieves an optimum globally minimum value posts 34 iterations. Thus, the gradient achieved is about 0.0087696 at epoch 34. Figure 4.9 presents the training state statistics and the corresponding confusion matrix is displayed in Fig. 4.10.

4.5.2   Training State

In a training state, a set of example data is used for learning by the NN, that is, to fit the parameters or weights of the classifier. This is the initial stage of the NN-based classification in which the neurons train themselves with the available data by fitting weights to the input layers.

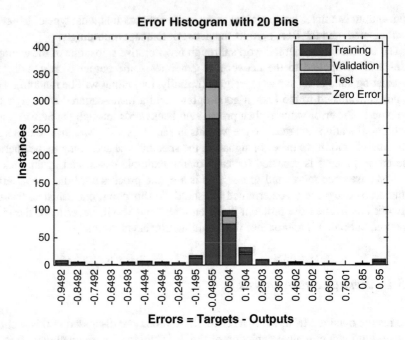

Fig. 4.8 Error histogram

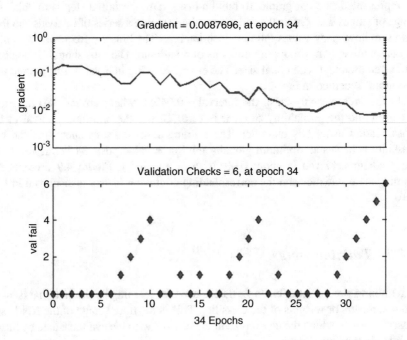

Fig. 4.9 Training state statistics

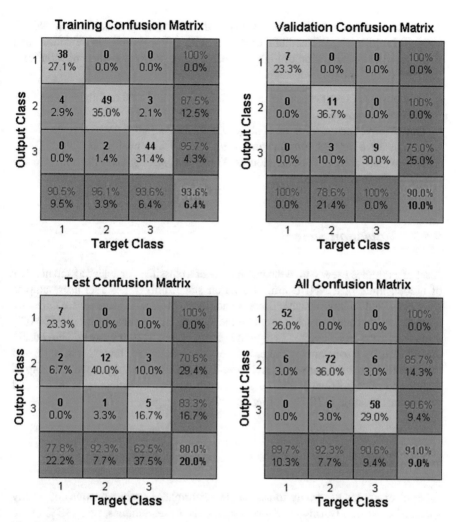

Fig. 4.10 Confusion matrix

During the training stage 38 user samples, i.e., 27.1% of the user samples, were correctly classified as low trustable users and only 4 (2.9%) user samples that were originally low trustable users were falsely classified as medium trustable users. Here, 49 (35%) users were correctly classified as medium trustable users by the system, and 4 (2.9%) users belongs to low trust users and 3 (2.1%) users belongs to high trust users which are wrongly classified as medium trustable users. 44 (31.4%) users who are categorized as highly trustable users were correctly classified, and only 2 users who belongs to medium trust category falsely classified as high trust category. It is important to note that none of low trust users were wrongly classified as high trust user and vice versa during the training state. 90.5%

correct classification of low trust users with 9.5% wrong classification, 96.1% correct classification of medium trust user with 3.9% wrong classification, and 93.6% correct classification of high trust users and 6.4% wrong classification were achieved during the training stage.

- True positive classification for low trust category is 100% and there is no false positive classification.
- True positive for medium trust category is 87.5% and false positive is 12.5%.
- True positive for high trust category is 95.7% and false positive is 4.3%.
- The overall correct classification is 93.6% and wrong classification is 6.4% during the training stage.

4.5.3   Validation State

A set of examples used to tune the parameters of a classifier for selecting the number of hidden units in a NN. During validation stage, only 7 (23.3%) user samples were classified as low trustable users. In addition, 11 (36.7%) users were correctly classified as medium trustable users by the system. In the same fashion, 9 (30.0%) users were correctly classified as high trust users, where 3 (10.0%) users were falsely classified as high trust among them. In brief, true positive for low trust is 100% and true positive for moderate trust is 100%. True positive for high trust category is 75.0% and false positive is 25.0%. The overall correct classification is 90.0% and wrong classification is 10.0% during the validation stage.

4.5.4   Testing State

A set of examples used only to assess the performance (generalization) of a fully specified classifier. During validation stage, only 7 user samples, i.e., 23.3% of the user samples, were classified as low trustable users. Consequently, 12 (40.0%) users were correctly classified as medium trustable users by the system. 5 (16.7%) users were correctly classified as high trust users, where 1 (3.3%) was falsely classified as high trust. True positive for low trust is 100% and true positive for moderate trust is 70.6%. True positive for high trust category is 83.3% and false positive is 16.7%. The overall correct classification is 80.0% and wrong classification is 20.0% during the test stage.

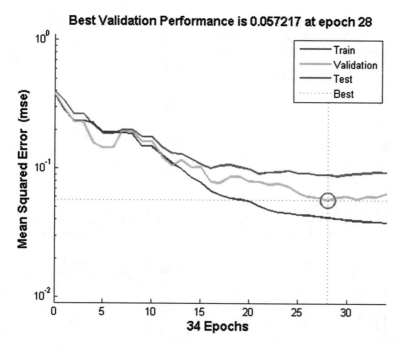

Fig. 4.11 Plot for overall performance analysis

4.5.5 Overall State

The NN confusion matrix is an important plot for the evaluation of performance of the NN-based classification as shown in Fig. 4.10. It is observed that the overall classification efficiency is 91% and false classification is 9% at the first instance of training. The level of classification is promising which can be improved substantially with a larger dataset with more number of training.

The overall performance analysis curve is highlighted in Fig. 4.11 and it shows that the validation achieves its peak without error at the epoch 28 with MSE 0.057217. The overall validation efficiency is always above the required level. The peak best performance occurred at epoch 28 with a MSE score of 0.057217. The state of testing and training is exponentially varying until the epoch 28, after that it is maintained steady state. The results are promising that the best performance occurred at the very less number of iterations. The ROC curve for training, test, and validation is presented in Fig. 4.12 and it shows the true positive rate for training state, testing state, and the overall true positive rates.

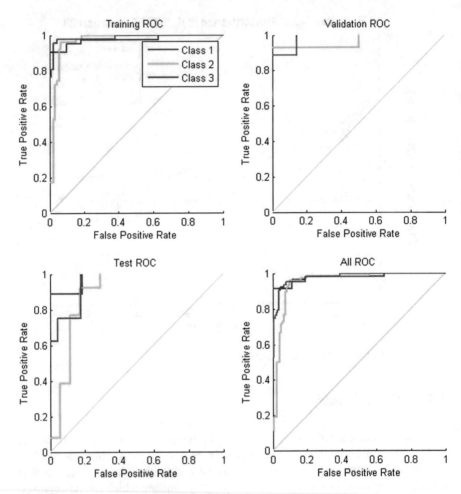

Fig. 4.12 ROC curve training, test, and validation

4.6   Conclusion

The trust issues and weakness in the role-based access control have been designed and developed in this chapter. Trust is a quality parameter in cloud computing in which NN-based BPA is applied to quantify the cloud services. The trust calculation method is presented with simple and reliable approach, which provides good scalability. The evaluation scheme takes behavioral history of the user that ensures trust issues in cloud with maximum accuracy. The proposed model effectively prevents the unauthorized behavior and unauthorized access. Through trust calculation, the CSP can select suitable services with dynamic trust values. The simulation result also shows that the proposed model helps the users to protect their data privacy from malicious attacks effectively.

References

1. J.V. Bibal Benifa, Dejey, Reinforcement learning-based proactive auto-scaler for resource provisioning in cloud environment, in *Mobile Networks and Applications*, (Springer, Berlin, 2018). https://doi.org/10.1007/s11036-018-0996-0
2. J.V. Bibal Benifa, Dejey, An auto-scaling framework for heterogeneous hadoop systems. Int. J. Coop. Inf. Syst. **26**(04), 1750004 (2017). https://doi.org/10.1142/S0218843017500046
3. F. Al-Turjman, H. Zahmatkesh, *An Overview of Security and Privacy in Smart Cities' IoT Communications* (Trans. Emerg. Telecomm. Technol., Hoboken, 2019). https://doi.org/10.1002/ett.3677
4. R. Kumar, R. Goyal, On cloud security requirements, threats, vulnerabilities and countermeasures: A survey. Comp. Sci. Rev **33**, 1–48 (2019)
5. S. Jabbar, S. Khalid, M. Latif, F. Al-Turjman, L. Mostarda, Cyber security threats detection in internet of things using deep learning approach. IEEE Access **7**(1), 124379–124389 (2019)
6. F. Al-Turjman, Intelligence and security in big 5G-oriented IoNT: An overview. Futur. Gener. Comp. Syst. **102**(1), 357–368 (2020)
7. X. Wang, L. Bai, Q. Yang, L. Wang, F. Jiang, A dual privacy-preservation scheme for cloud-based eHealth systems. J. Inform. Secur. Appl. **47**, 132–138 (2019)
8. S. Alabady, F. Al-Turjman, S. Din, A novel security model for cooperative virtual networks in the IoT era. Int. J. Parallel Prog. (2018). https://doi.org/10.1007/s10766-018-0580-z
9. F. Al-Turjman, Price-based data delivery framework for dynamic and pervasive IoT. Pervas. Mob. Comp. J. **42**, 299–316 (2017)
10. S. M. Khan, K. W. Hamlen. *Hatman: Intra-cloud Trust Management for Hadoop*, in IEEE Fifth International Conference on Cloud Computing, (Honolulu, HI, 2012), pp. 494–501. doi: https://doi.org/10.1109/CLOUD.2012.64
11. F. Yu, Y.-w. Wan, R.-h. Tsaih, Quantitative quality estimation of cloud-based streaming services. Comput. Commun. **125**, 24–37 (2018)
12. Y. Ruan, A. Durresi, A trust management framework for clouds. Comput. Commun. **144**(15), 124–131 (2019)
13. J. Prufer, Trusting privacy in the cloud. Inf. Econ. Policy **45**, 52–67 (2018)
14. G. Fortino, F. Messina, D. Rosaci, G.M.L. Sarne, Using trust and local reputation for group formation in the cloud of things. Futur. Gener. Comput. Syst. **89**, 804–815 (2018)
15. A. Silva, K. Silva, A. Rocha, F. Queiroz, Calculating the trust of providers through the construction weighted Sec-SLA. Futur. Gener. Comput. Syst. **97**, 873–886 (2019)
16. M. Tang, X. Dai, J. Liu, J. Chen, Towards a trust evaluation middleware for cloud service selection. Future Gen. Comput. Syst. https://doi.org/10.1016/j.future.2016.01.009
17. Y. Wang, J. Wen, X. Wang, B. Tao, W. Zhou, A cloud service trust evaluation model based on combining weights and gray correlation analysis. Secur. Commun. Netw. **2437062**, 11 (2019). https://doi.org/10.1155/2019/2437062
18. V. P, J. T, Multifaceted trust management framework based on a trust level agreement in a collaborative cloud. Comp. Electr. Eng. **59**, 110–125 (2017)
19. S. Singh, J. Sidhu, Compliance-based multi-dimensional trust evaluation system for determining trustworthiness of cloud service providers. Futur. Gener. Comput. Syst. **67**, 109–132
20. J. Kumar, A.K. Singh, Workload prediction in cloud using artificial neural network and adaptive differential evolution. Futur. Gener. Comput. Syst. **81**, 41–52 (2018)
21. F. Xu, C. Pun, H. Li, Y. Zhang, Y. Song, H. Gao, Training feed-forward artificial neural networks with a modified artificial bee colony algorithm. Neurocomputing (2019)
22. P.G. Brodrick, A.B. Davies, G.P. Asner, Uncovering ecological patterns with convolutional neural networks. Trends Ecol. Evol. **34**(8) (2019)
23. S. De Smet, D.J. Scheeres, Identifying heteroclinic connections using artificial neural networks. Acta Astronaut. **161**, 192–199 (2019)
24. L. Zhang, H. Li, X.-G. Kong, Evolving feed forward artificial neural networks using atwo-stage approach. Neurocomputing (2019)

25. I.E. Livieris, P. Pintelas, An adaptive nonmonotone active set – Weight constrained neural network training algorithm. Neurocomputing (2019)
26. T. Chen, Y.-C. Wang, A nonlinearly normalized back propagation network and cloud computing approach for determining cycle time allowance during wafer fabrication. Robot. Comput. Integr. Manuf. **45**, 144–156 (2017)
27. Amazon data set, https://archive.ics.uci.edu/ml/datasets/Amazon+Access+Samples, Accessed 15 July 2019
28. J. V. Bibal Benifa, Dejey. A Hybrid Auto-Scaler for resource scaling in cloud environment, J. Parall. Distr. Comp. (2018) https://doi.org/10.1016/j.jpdc.2018.04.016

Chapter 5
A Survey on Technologies and Challenges of LPWA for Narrowband IoT

Gautami Alagarsamy, J. Shanthini, and G. Naveen Balaji

5.1 Introduction

Mobile IoT refers to the 3GPP–LPWA standards which are classified as per licensed spectrum for NB-IoT and LTE-M. Narrowband IoT (NB-IoT) is a low-power wide-area network (LPWAN) 5G radio technology standard developed by 3GPP under release 13 to enhance existing wireless communication system. NB-IoT is an upcoming technology which is more appealing than other technologies in the LTE markets [1]. NB-IoT simply connects devices existing in mobile networks with high security structure. The NB-IoT is used to enable a wide range of cellular devices with low energy consumption, lower latency and reuse of LTE base with extended coverage. NB-IoT improves network densification and efficiency over the available spectrum. It offers deployment flexibility to support low-end applications.

Many IoT applications will function from small batteries or maybe from a harvested energy for a minimum part of the time, and as a consequence have a very strict power consumption price range. System-on-Chip (SoC) designers targeting the IoT market have specific challenges in turning in the developing set of capabilities required by the marketplace and preserving the low strength demanded by way of the utility. Often the device architecture requests to a utility processor degree of performance for executing superior device functions while retaining the power profile of an 8-bit microcontroller-primarily based device. The potential to configure a processor to gain those apparently conflicting goals is crucial. This article describes techniques and alternatives to lessen system energy via processor selection and configuration.

There are a couple of masterminded or persistent gigantic scale associations of the IoT to enable better organization of urban zones and structures. For example,

G. Alagarsamy · J. Shanthini · G. Naveen Balaji (✉)
SNS College of Technology, Coimbatore, Tamil Nadu, India

© Springer Nature Switzerland AG 2020
F. Al-Turjman (ed.), *Trends in Cloud-based IoT*, EAI/Springer Innovations
in Communication and Computing, https://doi.org/10.1007/978-3-030-40037-8_5

Songdo, South Korea, the first of its sort totally arranged and wired clever city, is gradually being worked, with about 70% of the business area completed as of June 2018. An incredible piece of the city is needed to be wired and robotized, with, for all intents and purposes, no human intervention.

Another application is at present encountering adventure in Santander, Spain. For this sending, two approaches have been grasped. This city of 180,000 inhabitants has quite recently watched 18,000 downloads of its city wireless application. The application is related with 10,000 sensors that enable organizations like halting the pursuit, natural checking, automated city plan, and that is just a glimpse of something larger. City setting information is used in this sending so as to benefit shippers through a radiance deal instrument subject to city lead that objectives growing the impact of each notice.

Various occurrences of gigantic scale associations in progress fuse the Sino-Singapore Guangzhou Knowledge City; tackle improving air and water quality, decreasing clatter defilement, and growing transportation adequacy in San Jose, California; and sharp traffic the administrators in western Singapore. Using its RPMA (Random Phase Multiple Access) developments, San Diego-based Ingenu has created a nation over open framework for low-move speed data transmissions using the proportionate unlicensed 2.4 GHz extend as Wi-Fi. Ingenu's "Machine Network" covers in abundance of 33% of the US people across more than 35 critical urban territories including San Diego and Dallas.

French association, Sigfox, began developing an ultra narrowband remote data mastermind in the San Francisco Bay Area in 2014, the principle business to achieve such a sending in the USA. It as such announced it would set up a whole of 4000 base stations to cover an aggregate of 30 urban territories in the USA prior to the completion of 2016, making it the greatest IoT orchestrate incorporation provider in the country thusly far. Cisco in like manner participates in sharp urban network adventures [8]. Cisco has started passing on progresses for Smart Wi-Fi, Smart Safety and Security, Smart Lighting, Smart Parking, Smart Transports, Smart Bus Stops, Smart Kiosks, Remote Expert for Government Services (REGS), and Smart Education in the 5 km region in the city of Vijayawada.

Another reason for an enormous sending is the one wrapped up by New York Waterways in New York City to interface all the city's vessels and have the alternative to screen them live each moment of consistently. The framework was organized and worked by Fluidmesh Networks, a Chicago-based association making remote frameworks for fundamental applications. The NYWW framework is at present giving consideration on the Hudson River, East River, and Upper New York Bay. With the remote framework set up, NY Waterway can accept accountability for its naval force and voyagers to such an extent that it was not effectively possible. New applications can be of ensuring security, imperativeness, naval forces, the administrators, propelled signage, open Wi-Fi, paperless ticketing and others.

Encompassing knowledge and autonomous control are not part of the primary thought of the Internet of Things. Encompassing understanding and self-overseeing control do not generally require Internet structures, either. Regardless, there is a move in inquisitive about (by associations, for instance, Intel) to organize the thoughts of the IoT and self-administering control, with beginning results towards this bearing considering objects as the primary purpose for self-overseeing IoT. A promising approach in this setting is significant stronghold acknowledging where an enormous bit of IoT structures give a dynamic and natural environment. Training a pro (i.e., IoT device) to continue acutely in such a circumstance cannot be tended to by customary AI computations, for instance, oversaw learning. By helping the learning approach, a learning master can recognize nature's state (e.g., identifying home temperature), perform exercises (e.g., turn HVAC on or off), and learn through the growing accumulated prizes it gets in the whole deal.

IoT knowledge can be offered at three levels: IoT devices, Edge/Fog centers, and cloud computing. The necessity for keen control and decision at each level depends upon the time affectability of the IoT application. For example, an independent vehicle's camera needs to make continuous block acknowledgment to keep up a vital good ways from a setback. This speedy essential initiative would not be possible through moving data from the vehicle to cloud cases and return the desires back to the vehicle [7]. Or maybe, all the action should be performed locally in the vehicle. Consolidating pushed AI figurings including significant learning into IoT devices is a working assessment region to make astute articles closer to this present reality. Plus, it is possible to get the most motivation out of IoT associations through looking at IoT data, removing disguised information, and predicting control decisions. A wide collection of AI procedures has been used in the IoT territory going from traditional methods, for instance, backslide, reinforce vector machine, and discretionary timberland, to bleeding edge ones, for instance, convolutional neural frameworks, LSTM, and variational autoencoder.

Later on, the Internet of Things modules may be a nondeterministic and open framework in which auto-created or sharp components (web organizations, SOA parts) and virtual objects (images) will be interoperable and prepared to act self-sufficiently (looking for after their own one of a kind targets or shared ones) depending upon the particular circumstance or conditions. Self-monitoring is conducted through the social means and considering the setting information similarly as the thing's ability to perceive changes in the earth (lacks affecting sensors) and present suitable control measures set up a critical research trend, evidently expected to offer legitimacy to the IoT development. Current IoT things and game plans in the business focus utilize a wide scope of advances to support such setting careful computerization, yet continuously present day sorts of knowledge are referenced to permit sensor units and savvy computerized physical structures to be sent in veritable environments [11].

5.1.1   Incorporate DSP and RISC Processors

IoT devices are defined with the resource of their functionality to take in or "sense" real international indicators, carry out operations at the related information, and communicate outcomes over a network, whether it is the net or neighborhood community. Most popular-cause RISC processors can approach the signals successfully; however committed DSPs can perform those responsibilities with higher energy performance and decreased latency. On the opportunity hand, RISC processors are perfect for shifting records and putting in conversation channels. Using separate independent processors is an option but adds value and board area to the device further to multiple improvement and debug environments and device. Key capabilities collectively with voice triggering, voice manage, speech playback, and inertial sensor processing, which might be needed in continuously on and coffee-power environments, leverage DSP instructions to carry out responsibilities which include filtering, fast Fourier transform (FFT), and interpolation while nonetheless meeting goals.

5.2 NB-IoT Architecture

NBIoT uses Evolved UMTS terrestrial radio access network (E-UTRAN) network architecture. The architecture core network depends on Evoved Packet Systems (EPS). It includes two optimizations for CIoT such as Control plane optimization and user plane optimization. The path of optimization is flexible to operate in uplink and downlink transmission modes. The packet data transmission with IP address takes place through PGW-packet data network Gateway and SGW-serving gateway whereas non-IP data transmission takes place through SCEF-Service Capability Exposure Function. The radio communication takes place between UE and MME through control plane. The architecture of NB-IoT is shown in Fig. 5.1.

5.2.1 NB-IoT Deployment Schemes

NB-IoT can be deployed in three different operation modes:

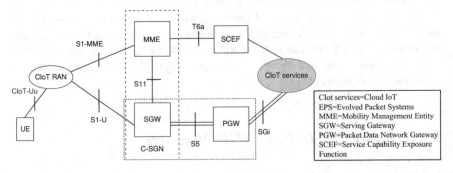

Fig. 5.1 Architecture of NB-IoT. *CIot services* cloud IoT, *EPS* evolved packet systems, *MME* mobility management entity, *SGW* serving gateway, *PGW* packet data network gateway, *SCEF* service capability exposure function

Fig. 5.2 NB-IoT deployment schemes

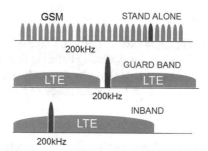

Table 5.1 Uplink and downlink transmission schemes of NB-IoT

Transmission schemes	Modulation schemes	Subcarrier spacing frequency	Slot duration
Uplink—single tone transmission	FDMA-GMSK	3.75 KHz	2 ms
Uplink—multi tone transmission	SCFDMA	15 KHz	0.5 ms
Downlink	OFDMA	15 KHz	0.5 ms

- Stand-alone mode
- Guard-band mode
- In-band mode

From Fig. 5.2 we can understand the deployment schemes that stand-alone mode utilizes a new stand-alone 200 KHz carrier and guard band mode utilizes unused source block within a LTE carrier guard band. Stand alone and guard band offer best performance at indoor coverage and power-saving modes under low cost [2]. Whereas in band mode utilizes single source block within a LTE carrier.

5.2.2 NB-IoT Transmission Schemes

NB IoT requires 180 KHz bandwidth for uplink and downlink data transmission. NB-IoT is integrated under LTE standard with a new air interface [3]. It uses licensed frequency bands which employs QPSK modulation. Based on the guard band and in band deployment the carriers are used for transmission.

For uplink SC-FDMA—Single carrier Frequency Division Multiple Access— is used. It supports both multi-tone and single tone transmission and for downlink orthogonal frequency division multiplexing is used as mentioned in Table 5.1. Table 5.1 shows the types of modulation used for uplink and downlink along with the frequency of subcarrier and frame slot duration.

The uplink F_{UL} and downlink F_{DL} frequency are defined as follows:

$$F_{DL} = F_{DL\ low} + 0.1\ (N_{DL} - N_{of\ FDL}) + 0.0025 * (2M_{DL} + 1)$$
$$F_{UL} = F_{UL\ low} + 0.1\ (N_{UL} - N_{of\ FUL}) + 0.0025 * (2M_{UL}),$$

where

$M_{UL/DL}$ = offset of NB-IoT channel,
$F_{DL/UL\ low}$ = operating band (uplink/downlink),
$N_{DL/UL}$ = E-UTRA absolute radio frequency number (downlink/uplink),
$N_{off\ DL/UL}$ = Minimum range of $N_{DL/UL}$.

5.3 Physical Features of NB-IoT

The physical features of NB-IoT are represented in Table 5.2. It represents the characteristics of NB-IoT in 5G wireless communication [4].

Table 5.2 Physical features of NB-IoT

Parameters	NB-IoT standards
Standardization	3GPP release 13
Spectrum	Licensed LTE band
Modulation	QPSK
Bandwidth	180 KHz
MCL coverage	164 dB
Downlink	OFDMA/3.75 carrier sub spacing
Uplink	FDMA with GMSK modulation /SC-FDMA
Link budget	150 dB
Peak data rate	DL:234.7 kbps; UL:204.8 kbps
Duplex operation	Half duplex
Power class	23 dBm
Communication	Message based
Energy efficiency	5–10 years battery life
Spectrum efficiency	Improved by band deployments
Coverage	Around 22 km(+20 dB GPRS)
Latency	<10 s
Maximum message/day	Unlimited
Connection density	1500 km^2
Interference immunity	Low
Peak current	120–300 mA
Sleep current	5 uA

Table 5.3 Comparison table of Sigfox, LoRa with NB-IoT standards

Parameters	Sigfox	LoRa	NB-IoT
Modulation	BPSK	CSS	QPSK
Frequency	Unlicensed ISM bands	Unlicensed ISM bands	Licensed LTE frequency bands
Bandwidth	100 Hz	250 kHz and 125 kHz	200 kHz
Maximum data rate	100 bps	50 kbps	200 kbps
Bidirectional	Limited/half-duplex	Yes/half-duplex	Yes/half-duplex
Maximum messages/day	140 (UL), 4 (DL)	Unlimited	Unlimited
Maximum payload length	12 bytes (UL), 8 bytes (DL)	243 bytes	1600 bytes
Range	10 km (urban), 40 km (rural)	5 km (urban), 20 km (rural)	1 km (urban), 10 km (rural)
Interference immunity	Very high	Very high	Low
Authentication and encryption	Not supported	Yes (AES 128 b)	Yes (LTE encryption)
Adaptive data rate	No	Yes	No
Handover	End-devices do not join a single base station	End-devices do not join a single base station	End-devices join a single base station
Localization	Yes (RSSI)	Yes (TDOA)	No (under specification)
Allow private network	No	Yes	No
Standardization	ETSI on the standardization of Sigfox-based network	LoRa-Alliance	3GPP

5.3.1   Technical Differences of NB-IoT Factors with Other LWPA Technologies

The technical differences of NBIoT with Sigfox and LoRa technologies [5] are compared and summarized in Table 5.3.

5.4   Advantages of NB-IoT

Advantages of NB-IoT are as follows:

- Optimized network architecture and end point density.
- It offers better scalability and QoS compared to other networks such as LoRa/Sigfox [10].
- Long battery life over 10 years.
- Bidirectional communication-uplink and downlink capability.

- Global standard in licensed spectrum-Release 13.
- Flexibility which can operated in 2G, 3G, and 4G without gateway.
- Improvised indoor coverage and propagation.
- Support maximized lower throughput devices under low cost.
- Low delay sensitivity and device complexity.
- LTE level security—Encryption and SIM-based authentication.
- Limited throughput.
- Deep penetration and better data rates.

5.5 Challenges of NB-IoT

NB-IoT faces challenges such as inadequate coverage and roaming problems which causes burden for service providers [6]. It is necessary for the network providers to concentrate on seamless mobility and better roaming agreements. The main challenges of NB-IoT among device manufacturers are as follows:

- Global footprint of Cellular Narrowband is inefficient in many countries.
- Cellular Narrowband is not deployed all over the world. NB-IoT is deployed over 31 countries with 50 networks. It is insufficient for IoT device manufacturers to provide NB-IoT with low-cost devices globally [9].
- Coverage area is restricted for the locations where LTE is already present.
- NB-IoT is implemented on LTE towers as stand-alone frequencies. To support the implementation guard-band and in-band deployments are explicitly needed. Since NB-IoT is used in LTE networks the coverage is not omnipresent. To provide better coverage the device manufacturers choose technologies such as GSM, UMTS, and LTE. This technique relies on expensive devices with prominent battery life.
- No NB-IoT roaming offered for commercial operations.
- There are no commercial operators for NB-IoT roaming as of today. IoT device manufacturers have to identify the roaming facilities for substantial business such as transport, fleet management, logistics, and cargo tracking.
- Poor choice for connectivity providers.
- Connectivity is an essential factor among service providers to design the embedded SIMs. Currently logistics and billing process have become increasingly complex worldwide whereas GSM, UMTS, and LTE technologies with global SIMs are available with a single billing channel.
- No redundancy coverage and availability of operator within a country.
- Across worldwide only few countries offer more than one narrowband services. Device manufacturer depends on the coverage area and opera ability. Multi IMSI-SIM Technologies such as GSM, UMTS, and LTE can provide a more reliable network.
- No seamless mobility and handover.

- NB-IoT does not support seamless mobility and handover due to power-saving reasons. It is not configured in real-time networks and applications.
- Limited support for SMS over Narrowband.
- SMS functionality is poorly deployed over NB-IoT. MT SMS are delivered only when the device is online whereas MO SMS can be sent via signaling channels under poor radio connections.

5.5.1 Adaption of Tightly Coupled Memory

The increasing demand for overall performance and processing abilities in IoT programs is riding a fashion to shift from 8-bit microcontroller tightly coupled embedded systems in the direction of 32-bit processor bus-based embedded systems. This shift negatively affects power and place of the system, which violates other key necessities of IoT products to be smaller and less expensive as they gain mass adoption. Tightly coupled extensions to 32-bit embedded processor systems can be leveraged to achieve all of those device goals concurrently by means of removing the less green bus infrastructure. The processor can get admission to memories and peripheral registers at once, lowering latency and required clock frequency, which reduces the quantity of energy required to perform the equal function.

Figure 5.3 compares a bus-based totally processor subsystem to a tightly coupled device processing sensor. The processor center accesses the auxiliary registers in a single cycle as opposed to at the least four cycles for the peripheral registers in a bus-based machine.

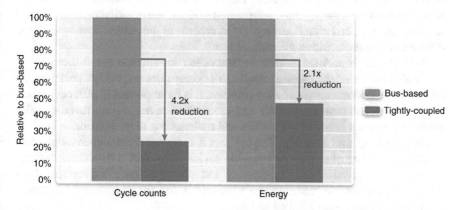

Fig. 5.3 Energy savings for sensor data in a tightly coupled system

5.5.2 Direct Memory Access

Another preference to reduce electricity in a processor device is to apply direct reminiscence get right of entry to direct memory access, which enables the peripherals to transport statistics without involvement of the CPU. To make sure, DMA must be highly optimized for the processor and application. Combining DMA with multibank reminiscence saves even extra strength because the internal DMA moves data inside and out of XY memory without impacting the processor pipeline.

5.5.3 Privacy and Data Security Challenges of IoT

However, there may be a huge amount of information constantly being generated and shared by using those IoT devices. In some of the IoT domains, security and privacy of information is the number one difficulty. For example, in a Smart healthcare use case, where user statistics and records is relatively privacy-touchy, the cloud-based IoT architecture increases critical records security and privacy troubles. In this structure, the statistics is hosted inside the cloud, which accommodates facts from individual devices, data aggregated from unique assets, and metadata associated with IoT entities. With those different sorts of IoT records, there are various facts security and privacy problems, which need to be addressed with in addition studies inside the cloud-based/cloud-enabled IoT structure.

Three main IoT information security and privacy challenges are data security, data ownership, and data privacy and sharing.

Data Security: In the cloud-enabled IoT architecture, it is far essential to ease consumer statistics and gadgets facts. Data get admission to manage fashions want to be advanced for securing static information saved at one of a kind factors, physical devices, gateway, and cloud, as properly as records in movement flowing between different components within the architecture.

Data Ownership: Another vital element of IoT information is statistics possession which has user data, gadgets, cloud carrier companies and the IoT packages. To deal with the information possession situation, it is far important to pick out one-of-a-kind sources and customers of IoT statistics, if the user needs control, the statistics ownership relation between IoT entities and statistics has to be defined.

Data Privacy and Sharing: The IoT fact is collected from numerous physical devices, consisting of sensors or wearable devices. This fact is then shared among several entities, physical gadgets, gateway, and more than one cloud services. Data privacy is an inevitable mission to be addressed in IoT for its persisted fulfillment inside the ever-developing linked global.

5.5.4 Models and Algorithms to Counter Attacks Based on Applications

To reduce vulnerability of attacks several models are followed based on applications.

- To provide flexible safety, a routing and monitoring protocol is designed with multi-variant tuples using Two-Fish (TF) symmetric key technique to find out and save the adversaries in the worldwide sensor community.
- Eligibility Weight Function (EWF) is designed to protect sensor nodes and it is hidden with the assist of complex symmetric key technique.
- A hybrid routing protocol is selected to be constructed through inheriting the houses of both Multipath Optimized Link State Routing (OLSR) and Ad hoc On-Demand Multipath Distance Vector (AOMDV) protocols.

5.5.5 Applications of NB-IoT

Some of the applications of NB-IoT are as follows:

- **Smart metering**—monitoring gas and water meters by small and regular data transmission over remote areas through deep penetration and excellent coverage.
- **Smart cities**—help to locate parking spaces, control street lighting and surveying environmental conditions such as littering.
- **Smart Building**—Security solutions, maintenance such as light, heat control through automated tasks.
- **Health Care**—Wearable IoT devices for health monitoring, people and animal tracking.
- **Fleet Management and Industrial applications.**

5.6 Conclusion

In this chapter we have discussed the architectural design, transmission, and deployment schemes of NB-IoT. From the detailed review, NB-IoT proves that it supports the 5G LPWA requirements and its standards. Compared to other technologies NB-IoT provides diverse deployment models and spectrum variations. It satisfies main applications such as critical communications, mobile IoT, and enhanced mobile broadband. The challenges and limitations can be concentrated in future and a better solution adapted in the IoT world.

References

1. Y.-P. Eric Wang, X. Lin, A. Adhikary, A. Grövlen, Y. Sui, Y. Blankenship, J. Bergman, S. Hazhir, A. Razaghi, Primer on 3GPP narrowband Internet of things (NB-IoT). IEEE Commun. Mag. **55**(3), 117–123 (2017)
2. A. Adhikary, Y.-P. Xingqin Lin, E. Wang, *Performance Evaluation of NB-IoT Coverage, IEEE 84th Vehicular Technology Conference* (IEEE, Montreal, 2016)
3. Z. Zhang, M. Zhang, F. Meng, Y. Qiao, S. Xu, S. Hour, A low-power wide-area network information monitoring system by combining NB-IoT and LoRa. IEEE Internet Things J. **6**, 590–598 (2019)
4. D. Deebak, E. Ever, F. Al-Turjman, Analyzing enhanced real time uplink scheduling algorithm in 3GPP LTE-advanced networks using multimedia systems. Trans. Emerg. Telecommun. **29**(10), e3443 (2018)
5. F. Al-Turjman, 5G-enabled devices and smart-spaces in social-IoT: An overview. Fut. Gen. Comp. Syst. **92**(1), 732–744 (2019)
6. F. Al-Turjman, I. Baali, Machine learning for wearable iot-based applications: a survey. Trans. Emerg. Telecommun. Technol, e3635 (2019). https://doi.org/10.1002/ett.3635
7. F. Al-Turjman, E. Ever, H. Zahmatkesh, Small cells in the forthcoming 5G/IoT: traffic modelling and deployment overview. IEEE Commun. Surv. Tutor. **21**(1), 28–65 (2019)
8. F. Al-Turjman, C. Altrjman, S. Din, A. Paul, Energy monitoring in IoT-based ad hoc networks: An overview. Comp. Electr. Eng. J. **76**, 133–142 (2019)
9. F. Al-Turjman, S. Alturjman, Context-sensitive access in industrial Internet of things (IIoT) healthcare applications. IEEE Trans. Ind. Inform. **14**(6), 2736–2744 (2018)
10. F. Al-Turjman, QoS–aware data delivery framework for safety-inspired multimedia in integrated vehicular-IoT. Comp. Commun. J. **121**, 33–43 (2018)
11. F. Al-Turjman, M. Abujubbeh, IoT-enabled smart grid via SM: An overview. Fut. Gen. Comp. Syst. **96**(1), 579–590 (2019)

Chapter 6
Framework for Realization of Green Smart Cities Through the Internet of Things (IoT)

Abhishek Kumar, Manju Payal, Pooja Dixit, and Jyotir Moy Chatterjee

6.1 Introduction

Now, there are two technologies developed, which is IoT and SC. Both technologies have caught the attention of both education and businesses. However, both types of technology consist of the same ideology. The IoT and SC both have apart genesis. These both technologies do not have apparent and brief definitions because it reflects their small history and prevalence.

By briefly investigating the genesis of both ideas, we got clear conceptions to know their potentials. However, the word 'IoT' was created in 1999. There are some techniques available that capable IoT like sensor networks appeared since the 1990s. Due to the decrease in the progress, processing and storage capacity of the sensor and cloud technology and the cost of sensor production, the deployment of sensors has enhanced in the last 5 years. In 2020, there will be 50 to 100 billion devices linked to the network which is predicted through the European Commission. In 2008, the quantity of things connected to the web had outperformed the quantity of individuals on worldwide which is shown in Fig. 6.1.

According to the definition, IoT permits people and things to use any network and service, anytime, anywhere, with anything and anyone. As we can see, the (IoT) is basically consumed through technological progress, not the requirements of

A. Kumar
Chitkara Institute of Engineering and Technology, Chitkara University, Rajpura, Punjab, India

M. Payal
Academic Hub, Ajmer, India

P. Dixit
Sophia Girl's College, Ajmer, India

J. M. Chatterjee (✉)
LBEF (APUTI), Kathmandu, Nepal

© Springer Nature Switzerland AG 2020
F. Al-Turjman (ed.), *Trends in Cloud-based IoT*, EAI/Springer Innovations
in Communication and Computing, https://doi.org/10.1007/978-3-030-40037-8_6

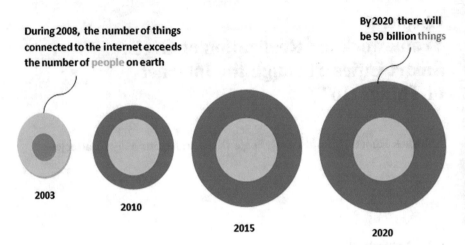

During 2008, the number of things connected to the internet exceeds the number of people on earth

By 2020 there will be 50 billion things

2003

2010

2015

2020

Fig. 6.1 Development of 'things' connected to the Internet

the applications or consumer. On the contrary, SC was created solving the modern cities related problems. The urban life has become a big challenge for both citizens and city administration because of rustic relocation and rural fixation towards urban areas. There are some critical issues available like traffic, water, education, waste, unemployment, health and crime management. These types of challenges are accurately and appropriately solved by the SC. The SC used ICT to solve these types of challenges. According to the definition, SC consists the six types of features which are as follows:

1. Smart People
2. Smart Economy
3. Smart Mobility
4. Smart Governance
5. Smart Environment
6. Smart Life

In Fig. 6.2, SC and IoT, whose origins are distinct? Both are enhanced away from one another for obtaining a general aim. We assume that SEaaS model consists of these two with many other technical and commercial models. Here, SEaaS refers to the 'Sensing as a service' miniature [1].

The target of SC applications is to increase the personal satisfaction of residents, enhance the use of efficient public and private assets, and decrease the nuisance, pollution and crime. SC is introduced about the particular needs for IoT system designs. Particularly, IoT systems will be utilized through customers with multiple and diversified SC applications and a large amount of data generated. It is quite different from the normal IoT system. It is designed to support special single-purpose application. For instance, GPS (Global Positioning System) finds information accumulation from smartphones and video images of passengers from

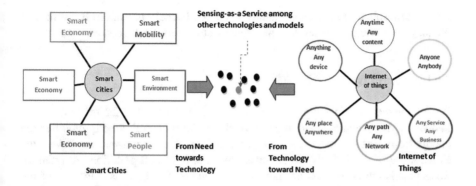

Fig. 6.2   The relationship between IOT and SC using sensing-as-a-service

city cameras and can be used in conjuction by a government in the city plan, the allocation and routing of buses and trains in public transport operators, logistics businesses in package distribution customization, and in planning passengers' self-travel.

Thus, information accumulated and utilized in the IoT systems may act as an object which can be a deal between information manufacturer, processors, vendors and consumer/users. In addition, like SC applications primarily objective at common customers, such as Governments and personal customers. Here, it is not of much necessity to explain IoT device layer details in applications like functionalities and structures. Mostly, the refined information and services are needed to be dispatched to customers.

In this task, we are primarily discussing the market-oriented concept of SC IoT systems with emerging information transfer and assets allocation methods. We have shown a data-oriented layered framework of SC IoT applications in Sect. 6.2.

We identified that a few distinctive connected work are of a socio-technical means that makes SC IoT systems totally apart from standard IoT from a "value of information" prospective in Sect. 6.3. Typical style concerns are reviewed in Sect. 6.4. Further, in Sect. 6.5, we have a tendency to propose a game theoretic market model for data commerce in smart-city IoT situations. Some numerical examinations which showed the smart-city partners' behaviours and edges of the knowledge commerce are displayed [2].

6.1.1   Introduction to Green Computing Technology

Green technology is again referred as sustainable technology. It is having a 'green' motive. Green is, of course, an allusion to nature, yet green technology, in normal, is one which accepts into account the long- and short-term effects creativeness has on the atmosphere. Green manufacture is an eco-friendly invention. It consists of the recycling, energy efficiency, renewable assets, safety, health concerns and so on.

An Instance of Green Technology

It contains the most legendary instance which is a solar cell & the method used is known as photovoltaic. Using the photovoltaic, the solar cell directly modifies the energy from light into electrical energy. By making electricity from solar power, the expenditure of fossil fuels, greenhouse gas emissions, and pollution are reduced. Solar panels are costly and are not much attractive. But there is some use of the solar panel like for inventions on the horizon which comprises the community solar group. It is renters to share the products of the solar panel. Modern spray-on photovoltaic film is used through the perovskites. Perovskites have the ability to convert window glass to solar collectors regularly.

Another easy invention available which may be considered green is a reusable water bottle. It is a healthy practice when drinking lots of water. It also decreases the plastic waste which is good for the atmosphere. Therefore, water bottle reusable method which you can replenish yourselves are eco-friendly, health-promoting and green [3].

It is also giving us game-changing manufacture in real terms. As an environmentally friendly customer, we definitely do not require to pursue our likes for devices. Now, there are two types of methods available for recycling plastic bottles, mercury-free screens and chassis. For mobile devices, batteries are much energy able than ever, which breaks the energy consumption of the house. It is also used for powerhouse applications like from speakers to laptops, using the energy conversion assets. And now that there are more eco-friendly car alternatives for hybrid electric vehicles than before, you can use particular GPS for calculating the most environmentally efficient path for your destination. It is not only that our manufacture is bigger than ever, but also that they are connected by the Iot technology and they are building their home in a customized ecosystem.

These types of manufacture can have a significant effect on our ecological footprint. It can only get benefit from the top of the line contract production corporates. EMS Companies, also known as electronic manufacturing services, comprehend the requirements of green technology developers who want to go fast for a business product in one go. EMS providers consist of the essential in-home specialization for supply mass manufacture of electronic equipment. It also consists of the EMS PCBA (printed circuit board Assembly). Mass capability allows EMS providers to offer high-quality and cost-effective production. There are many types of manufacturing businesses available which provide an integrated supply chain and data management services which enable it to control the challenges of low-volume and high-mix manufacture. Its EMS are done in many places, and it is offering the scalability for manufacture in several countries. For the quick developing and innovatory sector of green technologies, EMS business is worthful. The smart, cities are made with the help of Green technologies which have three types of pillars available, namely, a planet, people and profit.

Stockholm, Zurich and Singapore by using smart technologies and keeping these three pillars balanced have reached the podium of the world's most sustainable cities. The environmental pillar is related to greenhouse gas emissions, green urban spaces, renewable energy share, city's energy consumption, air pollution and drinking water. Zurich, for instance, has utilized high scaling techniques to recognize the city's most heated fields and to create methods for battling urban warming. Using the self-monitored power structures, smart grids connect the solar panels to LED streetlights that respond consequently when some portion of the framework is down. The hospitals consist of the own necessity solar power reserve. Streetlamps know when to diminish amid moderate hours. These abilities are only a little piece of a synergetic, accomplished, and it creates probably through the EMS corporates who give contract made electronics. EMS providers include the smart manufactory, small SC where sustainability and productivity are accessories to provide the efficient electronic manufacturing services and top-shelf. In the ecological pillar of smart and sustainable cities, Green technologies play an important role. This is a static to seek EMS to reach the mass production of their green innovations. The EMS for green technologies and SC have one specialist, namely, Asteelflash [4]. There are some points available for understanding SC well which are as follows.

6.1.2 From Digital City and IC to SC

There are many different words available which are used to describe whole ICT to increase the performance and capabilities of whole recent cities. There are three types of concepts available which are different from each other, namely, digital cities, IC and SC. These three concepts are used to describe the issues of research and development of the city operated by ICT. The alteration in words utilized to explain ICT-powered cities demonstrates the natural development of the strategies to increase the attributes of the life of the city. One digital city indicates to the digitalization of the city. It incorporates the information technologies, visualization and networking to reach the assets, population, economic, environment and social data. It is used for fulfilling the needs of the government, citizens, and businesses using the combined computing infrastructure and communication. The main objective of this digital city is sharing networks and information. One of the main instances of recent digital cities is Chicago. Chicago has manufactured its digital metropolis with the greatest network. An IC can be explained like a city outfitted with ICT framework. A SC might be treated between a learning society and a computerized city. It is a area wherein the neighbourhood arrangement of development is expanded through the intuitive instruments, implanted frameworks and computerized coordinated effort spaces. The purpose of it is to change lives and tasks inside this territory into important and basics instead of incremental methods. Technically, the characteristics of the city are described by the digital city, yet the

SC consists of the government and human facts. SC also consists of the technology. Therefore, a digital city isn't inevitably smart. But, first of all, the SC should be digitalized. It is more focused on technical grounds and has clear limitations. The SC consists of both types of sustainability and technology. According to some researchers and practitioners, an IC is usually as an alternative for a SC. Usually, people don't understand about the differences between IC and SC. Why the word SC is becoming more prevalent and admitted in solving recent urbanization challenges is explained by Nam and Pardo. The word 'Smartness' is focused on a consumer perspective. This word is superior to the more elitist word 'intelligent'. The meaning of 'Smart' is the capability of self-adapting and supplying services and customized interfaces to customer requirements. It is much eco-friendly than 'intelligent', meaning that it is responsive to quick brains and responses. In essence, we utilize the accompanying improvement as our method of distinguishing between conditions. The communications, procedures and information are as a whole digitized which is called the digital city. The IC is a digital city in which there is a layer of intelligence. This layer can create hive level decisions. These decisions are dependent on a phase of AI. The full form of AI is artificial intelligence. An SC is an IC where the application is cantered on customer experience and actual use [5].

6.2   Definition of an SC

To get knowledge about the scope and content of the SC it is important for them to understand the definition of the SC. As proved in Sect. 6.1, a solid definition of the SC is now emerging. The stakeholders through several definitions gave some static points. The definition of the SC is formalized difficult, since a city's smartness may be complex as a whole administration process describes the single function which is given to a decided exertion of a government process [6].

6.2.1   Review of Definitions

The SC definitions have been conducted in many surveys. Yet, mostly, there are so many different definitions about the SC which are collected and explained through surveys. A clear classification is lacking for SC definitions from several perspectives. There are four types of approaches available for the study and break down the meanings of an SC in this section which is below

1. Technical infrastructure.
2. Application domain.
3. System integration.
4. Data processing.

Technical Infrastructure

It is the first description given by Harrison et al., explaining SC as an IC, instrumented and interconnected. The connection between the physical, business infrastructures and social of a city is emphasized by this definition.

In the literature survey of smart city to be integrated into smart computing technologies which applicable to infrastructure components. These infrastructure components consisting of the software, hardware and network technologies are collected. It is considered by Washburn et al. that in SC whole structures must be manufactured, designed and maintained by using advanced sensors, integrated content, networks, and electronics, with databases and tracking. According to Bowerman et al. decision-makers with computerized systems interfere with algorithms. It is created for whole essential amenities which are water, electricity, transport and so on. In the SC, the communication protocol on network elements is used for representing the acquisition and transmission of data [7].

Application Domain

Giffinger and Gudrun define an SC from the perspective of domain application [8] which consists of the six types of smart attributes that are used for the definition and evaluation of SC which is given as follows:

1. People
2. Economy
3. Mobility
4. Governance
5. Environment
6. Life

An SC should utilize infrastructure and smart computing technologies. These technologies are used to create the city services more interconnected, smart and effective, which consist of the education, administration, public safety, health, actual resources, transportation, and availability [9]. Washburn et al. explained a SC as a solution which includes the water, electricity and gas utilization. It also includes public safety, mobility, heating, waste management, and cooling system. This is explained by [9].

System Integration

A SC's field application and technical infrastructure may be recognized as a cluster of integrated systems, interconnected and subsystems. A few scientists have tried to characterize a SC according to this perception. A SC explained as a natural mix of frameworks and their interconnection systems to create a system of systems smarter are defined in [10]. For a city to be smart, combination of city frameworks is fundamental so as to give adaptability and access to continuous data for creation and conveyance of proficient administrations. According to Javidroozi et al., for creating a smart city, it is supposed that the unification of city systems is important, in order to supply ductility and access to actual time data for manufacturing and distribution of efficient services [11].

Data Processing

Harrison et al. [7] have described that a SC according to data processing facilitates the acquisition and integration of existing actual global data through the sensor's resources. Interconnection gives the permission for the data collected from the assets to be unified across systems, numerous processes, industries or value chains. Lastly, the meaning of intelligence is that the data processing provides recent insight into running actions and decisions, which may display a concrete combined value. SC are integrated and interpreted with additional data processing technologies with the data to obtain smart services [6].

6.3 Literature Review

Wantmure and Dhanawade [12] are progressing slowly in Indian cities and cities around the globe. With the planned infrastructure, it is not an unexpected decision. The plan of SC has been recommended in an arranged city, with such effects that all actions did in the city are controlled and managed through technology. IoT is a rising modernization in the IT global that can be investigated to its pinnacle to accomplish the target of structuring a SC. A building alone is not sufficient, however, to keep up and continue their character. The respectability and credibility are other works to be handled and executed. There are a few difficulties in creating a SC in India, as there are a few certain and express hindrances that must be stood up to. A brilliant city model isn't an answer on the grounds that every city is exceptional in its reality. Nevertheless, a model advancement is essential through having a coherent structure utilizing for SC utilizing IoT.

Ahmed et al. [13] have given a framework of the subject that concentrates to its present status and prediction about the essential tasks that will play later on and clarify analysis of big data into SC and discussions about their probable responsibilities in alteration to our way of life and, lastly, it looks at the probability of this upcoming technique which can violate our privacy and deceives.

While huge information can give huge value, yet in addition exhibits a critical hazard to our own security and protection. The data isn't only more data, it is a product that can be purchased and sold through governments, partnerships and people. This chapter is a call to shoppers to think about how the enormous information gathered on them and used to tie their opportunity is dependent upon them to choose the dimension of information that is agreeable to share. To put it plainly, the advantages will exceed the dangers when it regards and ensures the rights and opportunities in just a society.

As per [14] smart city advances are advanced as a compelling method to counter and oversee vulnerability and hazard right now, yet they incomprehensibly make new dangers, including making city foundation and administrations insecure, brittle, and open to broad types of vandalism, disturbance and criminal misuse. This chapter is based on the first, and from a movement of models, ordered how urban zones are equipped with cutting edge devices and system that generates 'gigantic data'. Such information safeguarding the sharp city battle grants consistent examination of city life and modern techniques for rural organization and gives the unrefined item to imagine and authorize logically capable, profitable, supportable, aggressive, beneficial, straightforward urban areas.

According to [15] SC is created with the integration of information and communication technology, which effectively grants urban infrastructure management, heavy data generation, sensor networks, digital infrastructure and IoT through social networking, biological websites, geospatial data and GPS. This set of resources is essential for the database to create SC with health, business, transport, effective management governance, energy and assets. Taxonomy of the sets of unstructured and structured data: In this, taxonomy integrated into the cloud atmosphere with suitable software, hardware and interfaces will be advantageous. It gets suitable information according to consumer's requirements. It will be relieved in the SC long-term management.

After the commercial revolution, pollutants like carbon dioxide have been put into the atmosphere by the human civilization. These types of pollutants make the atmosphere unhealthy for the people. The Word of 'Global Village' was established in the 1920s. In those days, one media that was available for data communications is radio. Yet, now the global is interconnected using the networks. This network creates the whole world, factually a world village, because information can be easily sent and received from one space to another space.

Most of the electric and electronic apparatus have the sensors which is used to look global around there. They cannot be intelligent themselves. Yet, they give us, in most matters, information about their surroundings and how power is being consumed. In most of the urban areas, there are numerous sensors in regular life, i.e. there are sufficiency of information but reduction of interpretation. Wisely to reduce its footprint on nature so that quit behind what we have constantly given for the forthcoming descent. The approach of the SC is the achievement of its vision. In this chapter, we introduced the methods for utilizing the green software to create them smarter [16].

In the growth of world population, there are some types of fundamental problems such as higher cost, environmental alteration and air contamination. This problem arises when there is progressive reduction of the energy assets. This is the major problem that future cities will have to face to outlive, transform into SC and focus on smart mobility and green buildings [17].

In 2016, the value of CO_2 was approximately 32 Gigatons per year. This value was more than 50% of that in 1990, which so far is more than 400 parts per million (which is not happening for 300 million years). It is the reason of enhancing carbon dioxide concentration level in the environment. 12.6 million people around the world die due to environmental pollution, which is equivalent to a quarter of the total deaths.

Water, air and soil contamination, compound presentation, atmosphere changes, and bright radiation add to the expanding of more than 100 diseases and well-being harms. Barometrical contamination is the fourth hazard factor for passing on a worldwide dimension and without a doubt the principal natural hazard factor for lungs and heart infections: over 5.5 million individuals bite the dust each year every-where throughout the world on account of air contamination, more than Finland, Slovakia and Sicily occupants. The monetary assets that the overall urban areas have routed to adjustment measures to atmosphere changes like defensive boundaries against inundations, stronger foundations and better depleting frameworks (around 0.22% of GDP for the created nations contrasted with 0.15% for the urban communities of creating nations) are as of now applicable. Taking a gander at this situation, urban areas must be prepared and fit for taking care of gigantic social and natural changes, turning into the support of the battle against an unnatural weather change and catalysing speculations and strategies arranged to supportability and productivity in a Smart vision. The issues of sound pollution & air pollution of cities is due to the atmosphere changes. In the EU, structures alone are in charge of 40% of the last vitality use, 36% of CO_2 emanations or more 40% of particulate matter out-flows (PM10 and PM2.5). Current versatility frameworks dependent on petroleum product, other than being in charge of above 25% of contaminating discharges, are inadmissible to the requirements of urban territories, making developments [18].

Al-Turjman et al. [19] feature principle utilizations of smart cities and address the significant protection and security problems in the design of the smart cities. Singh and Al-Turjman [20] presented Cognitive Information-Centric sensor networks (CICSN), a worldview of WSNs in which sensory data is distinguished utilizing attribute-value pairs and components of insight are utilized to convey information to sink with client wanted quality of information (QoI) is presented. Dormancy and unwavering quality are recognized as attributes which affect the QoI saw by the client. With a utilization case investigation, it is shown how the CICSN can give client wanted QoI on the conveyed information. Ulusar et al. [21] presented some open-source instruments for taking care of tremendous measures of information that can be utilized to make answers for smart cities and examined some open research issues and empowering advancements, for example, energy consumption models, heterogeneous networks and security. Bhuvaneswari et al. [22] showed that maximum likelihood estimation which dependent on desired amplification is utilized for the parameter estimation approach; the assessed parameters are utilized for the training and testing of restorative pictures for ordinariness and anomaly. Başer [23] intends to inspect the substance of smart travel industry goal and its connection with SC tending for Antalya as a possibility for a smart travel industry goal.

6.4   How Green Software Makes a City Smart

SC are the cities which consumed technology to create diurnal life much efficiently, sustainable, intelligent and transparent. The SC relies on the utility of smart technology, the IoT and GIS software to refer to large collected big data.

6.4.1   Smart Operating Systems (SOS)

This is the OS which binds with the whole elements of the computer. The interface among hardware and the consumer is provided by the traditional OS. The OS is capable of interfacing among consumer, hardware, network and application software. The smart OS which we offer will be capable of communicating with the smart network. It is apart from the web or the Internet. It is used to optimize the performance of computers. The smart OS is constantly associated with the smart network, thus interacting the network with the interconnected computers and whole processing broker and the data centres. This uses the mass intelligence and capacity of the whole computer system into the SC. It should be noted that cloud computing architecture is fully apart from SC. The user is a dumb terminal in cloud computing. It is fully adverse in this matter. The computer is able to making autonomous calculations in this architecture. This is the only network for communicating with the network for integrated the data to reduce the functioning of the PC which depended on present energy/power manufacture and consumption phase (demand-supply relationship and environment friendly depreciation of power use) in the traffic and grid. Negotiation is necessary among other benefits in the network. The prevalence of it will be understandable when we make a picture of a city that reduces energy supply; only the computers reduce its efficiency, which is inevitable, to minimize the consumption of power or switch on battery energy. The smart OS utilizes the actual time data from the smart network. This data is used to start or end software applications which are based on fluctuations in power supply. This data is also used for obtaining target or reducing energy consumption. In this manner, the gross power consumed through the computers can be minimized numerous. The impact of power used through the computer isn't instantly clear because a personal computer doesn't use too much power. Yet the impact is accumulative. The meaning of this is that there so many computers available now and further a little improvement on their power used will rescue a large quantity of power.

   In some developed countries, 50% of computers are left overnight as shown in the research. This results in the predicted yearly power waste of 28.8 billion kWh and the economic costs of 2.8 billion dollars. A SC is an interconnected system. It can realize the power required and reduce the misuse of power. A smart OS will connect SC very well because the computers can conduct as sensors making themselves more conscious of the whole system. The smart OS will connect to SC data centres. It can ahead consist of the global urban virtual network of SC. It will connect the

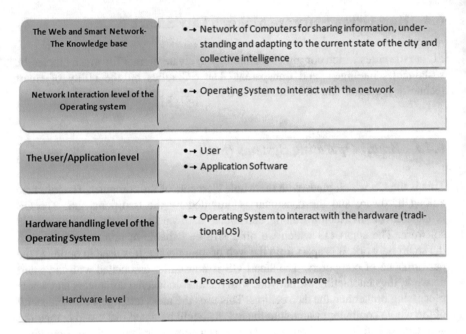

Fig. 6.3 Proposed architecture of SOS

upload data to power circulation and present power supply scenario. Looking at this, we present an architecture of a smart OS which is shown in Figs. 6.3 and 6.4 [24].

6.4.2 Inter-Software Interaction

The SOFTWARE utilization functioning on the common system are generally in discontinuity to one another. It could be seen like information privacy. However whole software applications have to communicate with the OS. This communication can be used to communicate with one another for software applications so that general requirements can be met. Let's say that a software application on a PC need data X, for example, to say about the latest weather updates. The common handheld computer also runs other software application. It requires the common information, then instead of getting it from the network, this should seek information from another application. It will minimize energy consumption and leads to processing time (more energy is wasted during more processing). In order to create its inter-software communication probable, the OS design should be such that the OS have a well understanding of every user that functions on it, i.e. it means that the OS will be intelligent and will be eligible for understanding about the types of information being downloaded from the network as well as knowing about the applications it is running. The OS is used to bring data and network. When the OS is conscious

Fig. 6.4 Integration of the smart network with present OS

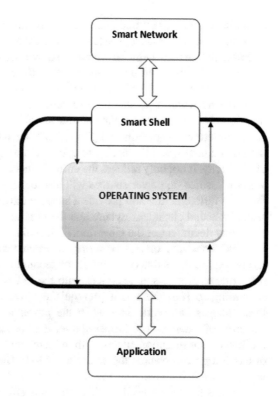

of the necessary information, then wherever this is applicable it can recapitulate the information. It is other facts of what OS could be eligible to make. That is, the smart OS through intelligent content awareness and complete understanding of what an application does would be capable to superiorly bridge the gap between the network and the application [24].

6.4.3   Utilization of Smart Power Grids

It is a network which connects power corporations to a list of intelligent devices. It maintains the tabs at electricity consumption within corporations and homes. It keeps the energy inconstancy phase which is based on the phase of current energy expenditure. The green software application can improve its conduct appropriately through this information. The energy expenditure is optimized, which is depended on the utility of energy, and therefore reduces energy dissipation. SC use electricity from the smart power grid. In the power supply, the overall consumption of power of the city can be adjusted. Hence, there will no unprecedented electricity failure. The aims of a SC are to utilize digital technology and communication. It uses the digital technology and communication for improving the applications of assets, for example, water, energy and roads, and infrastructure. It is also used to improve transportation, governance, waste management and health care [24].

From the energy point of view, a SC will be capable to optimize the city's power consumption through being capable to record the actual time data related to various commercial, residential and industrial locations. A SC is furnished with a smart grid which provides the different types of the facility for collection and transfers electric concerned data across the city. It is released of whole problems [25]. Therefore, the person living in SC will have essential control over its power usages and finally will be capable to optimize the expenditure transacted on the bills of electricity. It not only relieves the curbing uncontrollable power consumption, but heavy pressures on power sources will also be reduced. The promoting cities offer the main infrastructure and provide a suitable attribute of life for their citizens in a sustainable and clean atmosphere and the uses of smart solutions are mentioned in the policy document of the Government of India.

A SC consisted of less or more an urban vision. It contains many types of recent requirements like communication technologies and the integration of several platforms of products and services with the IoT to normally managing townships. According to [26] there is a prerequisite power for the whole of this because these devices can work only when the power supply is efficient. For obtaining the level of smart power infrastructure, it is necessary to be capable to monitor the power consumption along with remote control [27]. It is place where the discussion grows towards the solutions of IoT. The first phase is the Smart Home which is used for changing the infrastructure. Then data that is produced from each home has to be used intelligently and effectively for predictive control of power which can support to save the abundance and wasteful power consumption on giving huge monetary saving as well as on city level. The objective of SC is to improve the latest technologies and provide preferable well quality of life through an automatic mechanism, in which it is necessary to supply a consistent power. The main focus of SC is dependent on the utility of cheap, reliable and sustained supply of electricity, which together with the robust transmission and distribution infrastructure development mandated as well as augmented production. It is a significant success factor in acquiring smart power [28, 29].

6.4.4 Feedback Data

For the SC, green software should deliver feedback information for better capacity. The green software depends on the buyer. It very well may be as a PC control which is important to act right then and there. With the utilization of energy information of thousands of PCs, about the SC Analysis focuses which is talked later. SC Analysis focuses can offer guidelines to PCs to make elective like power sparing mode or keep running on battery control when the power is exhausted. Again, the OS should assume an exceptionally basic job here in light of the fact that this feedback signals can't be sent through various application software. Henceforth, we look that there is a smart OS in the focal point of the green software. So as to comprehend the present condition of the reaction network, different factors, for example, normal download and transfer speed, can be remembered. All things considered, the feedback will go to SC's Network Processing Centre [24].

6.4.5   Specialized Green Software

More precisely green software for SC. There are few discoveries as follows:

- **A Smart traffic system:** It can connect to the cars for actual time in information streaming (processors to work on the information sent in it). That is, when there is a traffic disorder in a specific area, it will have to be averted through the detouring. It will save time and, in this manner, save power. The traffic system of SC should be capable to advice the car driver to drive at optimal speed (on the basis of the car) which will reduce the gaseous emissions. In fact, it can also connect to smart networks which are explained in Sect. 6.4. In this manner, it may be a portion of the whole smart system.
- **Smart Street lighting:** It is a system which is used for switching off street lights in small roads when there is no one in the road. This cannot be suitable for major roads, but this will be possible to do surveillance (computer night vision) for the roads, which will automatically switch to the light when anyone is in the vicinity, otherwise, it is closed Will go. At the same time, it will likewise be connected to the smart system which will enable additional energy to be stretched out to energy lacking spots.
- **Smart transportation**: A city is a smart transport application which runs on the handheld computer through the green software. The green software provides the present state of the public transport, the distance of the user's place, regardless of the seats available there. This system will help the user to deliver the required time to reach the user, save energy through the advertising public vehicle and furthermore chip away at the example of travellers, for example, the vehicle recurrence might be limited in number around evening time [24].

6.5   Architecture of Multi-level SC

With the assistance of remote sensor systems and present-day remote advances, in urban social orders, we accept the eventual fate of Smart Cities frameworks which provide the powerful, flexible and intelligent support for individuals living.

As showed up in Fig. 6.1 we presented about a SC design which is an enlargement of 10, which was limited to the vehicle space so to speak through fusing remote sensor systems and available remote correspondence points of interest, the going with investigation thoughts are jogged on: (1) on-going irregular state setting

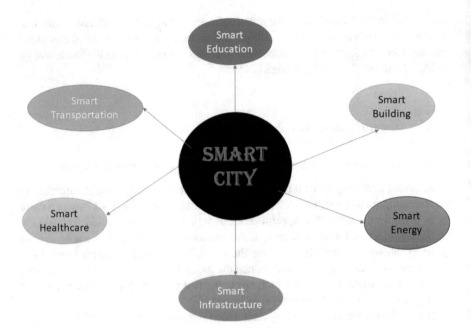

Fig. 6.5 SC components

carefully changed organizations; (2) upgraded living conditions and (3) improved utilization of the open properties [30].

As seen in Fig. 6.5, we visualize the fundamental components of the SC architecture to be the smart environment, smart health, smart security, smart energy, smart office, smart administration, smart industries, residential buildings and smart transport. The sensor nodes sent in every SC area give the essential information assets to heterogeneous data generation. Data produced using the sensor hubs are gathered utilizing the current communication services. For instance, the utilization of satellite network for GPS gadgets, mobile phones, for example, GSM/3G/4G for PDAs and the utilization of the web for PCs and another types route gadgets for crude information accumulation. The gathered data are then analysed and processed utilizing semantic web innovations and Dempster-Shafer rules. The emphasis is on expanding the design on a cloud platform for being utilized as a SaaS. The introduced structure can support Alzheimer's sufferers and older individuals with their everyday exercises, for instance, through alerting cautions and monitions clients on the off when they not be able to remember, or are incapable to finished, day by day living exercises. The framework will likewise serve as an intelligent stage for individuals living in a smart society.

Through joining information through various SC areas, this framework will support in helping individuals in a smart way, for instance, controlling a driver to take other path if there should arise an occurrence of street blockage, alarming heart patients in circumstances where their heart rate is surpassing an edge limit while playing out an action, helping individuals with cautions and admonitions for their household things, for example, sending alarms for purchasing food things through a Smart Fridge.

The execution of the framework will follow the phase laid out underneath. First of all, raw data is gathered and processed to creating web consumption. The information is changed over onto a typical format, which is further semantically enhanced with OWL ideas depending on the learning of space specialists. At a similar dimension, the gathered information is processed through the Dempster-Shafer principles to manage the vulnerability parts of the semantic model. The principles are to distinguish action and cognize about recent principles that are controlling an action. The recent principles learned at this dimension will be utilized in explaining the information of the semantic structure. A similar methodology will be utilized in explaining about the modified services that will give input to the end clients (natives) as alarms and admonitions as referenced in Level 4 of the SC architecture.

6.5.1 Architecture of Multi-level SC

As seen in Fig. 6.5, sensors structure the essential assets of information generation. The raw data discovered through the sensor node is swapped to Phase 1 of the SC design through communication services to do additional information processing. A definite depiction of every level is clarified beneath.

Phase 1: Information Collection

In this dimension, crude data accumulated through sensors is saved ahead processing. There are few formats available where heterogeneous information is gathered are tweets, csv, text messages and database schemas. The gathered data are then processed through semantic web advances so as to change over them into a typical format. The following dimension depicts the phase consumed in the transformation of information into a typical format.

Phase 2: Information Processing

Information gathered through the data aggregation stage is consolidated prior to transference, investigation and mix in the ahead dimensions through the semantic web developments. The principal motive of this dimension is to change over the accumulated heterogeneous data into a run of the complex position, for instance, RDF, alluded as the Resource Description Framework. RDF11 is a broadly perceived way to deal with interchange data over the Internet and this empowers heterogeneous information interchanging and coordination for various SC spaces. RDF in like manner help in clarifying metadata about the benefits on the Internet. Particular programming could then have the option to utilize RDF data for astute exercises. Pre-prepared RDF data delivered at this measurement will be abused through the semantic learning rules in the accompanying measurement for anomalous state reference cognizant information recuperation.

Phase 3: Information Integration and Reasoning

Semantic web advances enable maltreatment of room unequivocal data reliant on the plans and associations among those plans. The plans used in this measurement are shortened underneath. OWL is used for appropriating the ontologies. Here, OWL is alluded to Web cosmology language. It is an RDF diagram that is manufactured through the ontologies and RDF. Its authorization is the portrayal of the individual/Plans reliant on the class. This also delivers two separate kinds of characteristics, which could be used to outline associations among different classifications, specifically the information traits and object qualities. At the point when data game plan is done, learning can be moreover best in class with space masters. Dempster-Shafer will be utilized here for development affirmation and adjusting despise rules in a specific space of talk. In this chapter, the Dempster-Shafer technique is expended for merging sensor information six through various SC spaces. Its philosophy will help in adjusting late data by means of question-able reasoning and accordingly help with achieving a vigilant splendid structure. SPARQL is an RDF question language which is used to recoup, request and control data set away in the RDF position. At the point when the total database is conveyed as RDF altogether expands, SPARQL enables the inquiry and recuperation of data in a comparable association. Thusly, this measurement rouses among low-level information mix. The new principles got in the midst of the method of extraction of unusual state setting information through rough sensor information could then have the option to be put away and used for structure up learning in the SC engineering.

Phase 4: Device Control and Warnings

Data from phase 3 could be consumed through numerous Internet applications for intelligent working. The implicit data can be consumed from numerous points of view, for example, messaging, input/output, alarms and alerts [30].

6.5.2   *Communication Services*

The communication medium supposes a significant job in completing the SC plan. Figure 6.6 demonstrates the present communication benefits that are used in a SC foundation: LTE (Long-Term Evolution), 3G (third generation), ZigBee, WiMAX (worldwide interoperability for microwave access), satellite communication, Wi-Fi (Wireless Fidelity) and CATV (digital TV). The key objective is to associate an extensive range of things (IoT's and sensors) which could help in creating the life of natives increasingly agreeable and more protected. An instance is produced through communication advantages in the home space for interfacing phone assets and PC using the internet. On account of the Government Section, cloud and communication organizations are merged to acquire a greater administration system. On account of the health area, communication advancements can be utilized to associate medication, health statistics and place of the sufferer through a remote place and accomplishes a Smart Health framework. Subsequently, with SC and communication advancements we can provide an increasingly protected and advantageous foundation for superior existing [31].

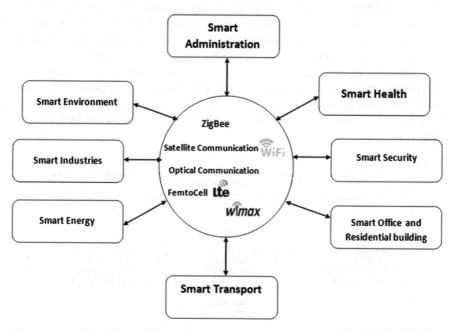

Fig. 6.6 Communication services

6.5.3 Customized Services

On account of the vehicle and health areas, through joining sensor information we can gauge the effect of driver health attributes on the driving situation. Joining health specification such as heart rate and blood pressure with vehicle status can assist the driver in measuring their constant health situation, which can help in making a sheltered situation for drivers. Likewise, utilizing vehicle area, vehicle speed and volume of traffic moving towards an intersection, we can help in better observing of vehicle status. On account of the human services area, data gathered through remote sensor organizes about patient well-being and movement can help the incapacitated individual. So also, by joining the home and condition spaces information, the impact of temperature on home exercises, for instance eating, washing, resting and cooking, can be educated. This can help in perceiving right action status, which thus can be a valuable consideration instrument for the old and individuals experiencing dementia. On account of nature and organization areas, low-level data gathered from the earth space, for example, temperature and water level, will help in determining abnormal state redid data.

At the point when abnormal state altered data (for example, flood, seismic tremor, timberland flame, avalanche, and other regular catastrophes) is joined with city organization administrations, it could help in sparing lives. Thus, on account of the modern segment, setting mindful administrations got through heterogeneous information combination will help in making a protected workplace for assembly line labourers. By persistent checking, recording and misusing the encompassing sensor data from various spaces, (for example, unsafe gas recognition, machine conditions, and specialists' well-being) in a mechanical situation, a superior, progressively beneficial and more secure condition for labourers can be made [31] (Fig. 6.7).

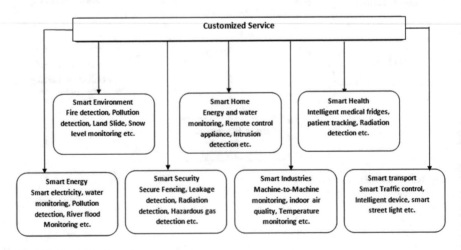

Fig. 6.7 Customized services

6.5.4   The Smart Urban System (SUS)

On the back of the SC, there is a framework involving green programming and computerized machine communication which gathers information from the whole city, examinations it and is being used for the proficient running of the city. We consider the SUS framework. The SUS runs with its objective to proficiency and eco-friendly way. Investigation of the present situation of the city by the SUS, assessed and estimated depending on the variables (such as on-going force request and supply, network blockage and load-distribution) will be known as the present condition of the city. The various parts of the SUS (which is seen in Fig underneath) are discussed beneath.

The Smart Network

This is a city-wide system which saves the present condition of the city and furthermore the power advancement strategies which is most proper for various classes of customers (such as PCs, handhelds) utilizing which the gadgets can manage their capacity use contingent upon the examination conducted by the server farms of the SUS. The intelligent system will be known as the essence of the SUS as it mirrors the adjustments in the framework and saves the prescribed activities. The information base of all investigation of information got from the SC is utilized to viably create the ideal power utilization levels. During the arrangement of continuous examinations of various parameters, the last condition of the city is assessed using the server farms and transferred to this system with suggestions on what could possibly be done on the present state regarding power utilization and the running of the gadgets.

The present condition of the city is observed at standard interims through the SC. It is done after examination of the information accumulated through the sensors that are available around the city. Smart gadgets such as telephones and PCs keep up a consistent (yet light) association with the Smart network. It empowers the gadgets to alter their conduct dependent on the ongoing condition of the city. Power outages and power deficiencies create over a moderately brief timeframe that every one of the gadgets needs to remain synchronized through the altering condition of the city.

The smart network is additionally associated with the Internet. Thusly, the gadgets associated with the smart network are the truth, likewise associated with the web. That is, each gadget endeavouring to associate with the web experiences the smart network that will pick the quickest path to the web server just as it could give measurable information to the examination of the system traffic for better network routing later on. As it were, the web would occur more intelligent with the SUS.

The SOS and Smart Assets

As referenced already, the smart OS will help the performance of smart gadgets with decreased energy utilization. The smart OS gauges the power needs of the PC and produces a power list in which the system plans the best power settings for the gadget. The smart OS follows up on the proposals of the smart network to the extent practicable.

The smart OS gives the review to the smart system dependent on its point of view. For instance, when the normal download speed of the system is uncommonly moderate, the review helps the information investigation focuses to assess the system paths and analyse issues in the network.

The client is associated with the application programming to play out his/her activity. The application programming interfaces with the smart network using the smart OS. In this way, the keen working framework is constantly mindful of how the applications are communicating with the system. This encourages data interchange (as referenced in the past segment) and good input to the framework. An OS shell is a product part which exhibits a UI to different OS functions and services. The smart shell will be introduced on the current OS. The shell would get the data with respect to asset utilization through the OS and computes the power utilization subtleties and sends it to the intelligent organize for assessment. The reaction through the smart network is additionally gotten through the smart shell and depending on this proposal the smart shell educates the working framework to change its usefulness. The smart shell transfers all other collaboration with the PC to the OS, that is, all other users of the PC is constant.

The OS computes a power list upon the applications as of now running and asset utilization. The power list is given to the smart network. In view of this power list, the smart network assesses the asset utilization and power necessities of the gadget. The power list is a 4-digit string (which could be broadened further) where every digit is a pointer to the particular power necessities of a class of uses. The power file is created as specified in Eq. (6.1).

$$\text{Power index} = 10^3 * \text{COS} + 10^2 * \text{CP1} + 10^1 * \text{CP2} + 10^0 * \text{CP3} \qquad (6.1)$$

Here, COS = indicates the requirements of the power.

Power necessities of the essential useful and related units of the framework (like the processor cooling equipment) evaluated through the OS. This is simply the power utilization of the OS itself. It has the most elevated need on the grounds that without it the PC won't work.

CP1 = It consists of the Power requirement (Pi) of the application. It consists of the topmost priority at the instant.

CP2 = It consists of the Power requirement (Pi) of the application. It consists of the average priority at the instant.

CP2 = It consists of the Power requirement (Pi) of the application. It consists of the lowermost priority at the instant.

Pi = Here, the Power requirement of an application-i is determined depending on variables such as normal CPU utilization, disk accesses, memory use, network action, application state (regardless of whether it is running out of sight or closer view), power state (whether keep running on battery or power).

Sensors

In a SC there are sensors around the city which estimates various parameters to create an assessment of its environment. The point is to make a distributed system of intelligent sensor hubs which can gauge numerous parameters for increasingly productive management of the city. In this way, there would be an immense number of sensors around the city. Whole information through the sensors would be sent because all things considered in its static structure to the key information analysis focus or the sensory information could be dissected at various dimensions, and thorough analysis, the measure of data sent through the sensors can be limited to save energy and bandwidth. The smart gadgets (telephones, PCs, tablets and so forth.) likewise go about as sensors using the review they send.

Data Analysis

It is the most fundamental piece of the SUS (here, the full form of SUS is System Usability Scale) so far since SC has a complete task. It depends on the amount of the information into the sensors that are translated for start guidance on how the gadgets could run based on a sensitivity of the city, here which has been investigated. This is based on how any kind of data set is right now being worked in the city. This is analysed to change over it into a vision which calls the current situation with the city and this is additionally dependent on the state. In this, the analysis is completed on upgrading the strategy of the gadget capacities.

There are four types of major subdivisions in the data analysis centre (this may be expanded as the city diffuse):

- Power/energy analysis: Smart power grids (supplies) and the energy use data are well read to be efficient to supply recommended energy through the guidelines. With smart grids to work with, the power management has been practically improved in SC. This part of the analysis centre also meets the energy utilization requirements through the renewable assets of power and reducing the use of non-renewable assets of power.
- Network analysis: In these parts controls the Smart networks and these functions. Depending on the feedback information into the devices and sensors within the network, the estimation may be created regarding the general functioning of the smart network. It is imperative to keep in mind that when the smart network defaults, the gadgets won't have any plan regarding the condition of the city and

hence a great part of the reason for this system would be beaten. In this manner, network analysis is a pivotal segment in the SUS.

- City-wide broad analysis: In this part, analyses about the information, for instance, road and air traffic, weather conditions, air pollution, analysis about the waste disposal and so forth. This part is used for the measurement of the state of every one of these cities. For instance, the heavy rain will enhance the likelihood of accidents declining, and therefore the speed limit of the car will be lower than the general value. This suggestion is gotten through smart cars through the smart network and gives the advise to the drivers for slowing down the car through each period which minimized the possibility of accidents.

- Statistical analysis: In this part, whole analysis completed through the 3 previously mentioned segments are stored as experience and statistics. The SUS cognizes as much as it works, through past experiences. The statistical data may mostly ahead subpart to achieve the potential solution to a complex problem. This part consists the adapting, machine learning, statistical analysis, and heuristics of data which is used for the better solution of recent issues than previous experiences. The data storage is control through the big data and data mining technologies [24].

There are some types of major issues available when altering a city as given below:

(a) Privacy issues
(b) Compatibility problems
(c) Educated and energy-conscious society
(d) Green policies
(e) Economy

Privacy Issues

The smart network has reached to the computer's power consumption requirements (like application-specific) in the proposed smart urban system. It doesn't have access to application data. Smart networks do not have direct access to applications and client data as this is done through the operating system. It is connected with the OS that shifts sensitive types of information. It is the only detail of power utilization.

Compatibility Problems

There is an immense number of PC gadgets right now running OS not good with the SUS. As we introduced in area IV, we can utilize the smart shell (which is seen in Fig. 6.5) to create the present OS agreeable with the SUS. In this manner, the SUS can improve the power utilization of current systems too. It is likewise to be noted that the issues presented through the incredible diversification in the presently running systems is taken care of through the smart shell.

Educated and Energy-Conscious Society

Even in spite of technological progress within the area of green software and energy conservation in a tailored urbanized global of SC, the success of the full sensible system model is predicated on whether or not folks are able to adopt this associate degreed settle for that an urban global SC is the solely answer to the climate crisis. This is ultimately in the hands of the folk of the globe through the SC will serve its objective, since unless there are consciousness and awareness, no change can happen. Since citizens are the main cause for the duration of city engagement and city policy. These two words can help cities assess and get their aims. It is especially appropriate in a global where citizens are disabled consumers instead of services. Now citizens become the "prosumers (manufacturer and consumer)". We can say that without smart folk, a global of SC cannot be created.

Green Policies

There must be create some national policies for software business and customers to increase green software produces compared to their non-green counterparts. It will affect the market to move completely into green system. This is necessary to do so since the public proposed above for the new green system will never change until it will straight benefit them financially in developing countries especially. This is because the general man generally doesn't look beyond the economy, so far green practices are withdrawing one seat.

Economy

We can understand about the importance of the role of the economy by the fact that populations from rural areas are driven by the need to create huge pressures on present cities and seek economic opportunities that demand the construction of novel urban centre like this, one of the reasons that the concept of SC is increasing in the near forthcoming due to the economic split among the urban and rural community. Like beforehand has been discussed, we proposed the smart system to optimize energy utilization and thus decrease the amount of money spent on energy. Yet there are other sections in the smart system which accept the processing, data collections, network maintenance, between one another. It enhances the cost of staying in SC. It should also be borne in mind that the enhancement in costs is becoming city-wide. It says, adding that the government may idea to create a budget to include the costs since this isn't an exceptional cost, yet with a cost SC will come. This can also supply income for the city while at the similar time reduce the cost of non-renewable assets of energy [24].

6.6 Conclusion

With the intensive development of the emerging Internet of Thing, we endorsement the SC using IOT. In this chapter we give a comprehensive overview of SC using IOT. We propose that, for strong definition of SC four key perspectives should be considered: technical infrastructure, application domain, system integration and data processing. We also suggest that how green software like smart operating software, information sharing software, IOT and GIS software help to make city smart. The related survey research also helps to provide primal support techniques and basic theory for different smart definition. Through SC architecture our aim is to focus on most vital area of SC environment. We also reviewed the life cycle of green IOT as well as required technologies to instate of green IOT system.

References

1. C. Perera, A. Zaslavsky, P. Christen, D. Georgakopoulos, Sensing as a service model for smart cities supported by internet of things. Trans. Emerg. Telecommun. Technol. **25**(1), 81–93 (2014)
2. Y. Zhang, Z. Xiong, D. Niyato, P. Wang, Z. Han. *Market-Oriented Information Trading in Internet of Things (IoT) for Smart Cities*. arXiv preprint arXiv:1806.05583. (2018)
3. The Many Reasons Green Technology Is Good Mary Bellis. https://www.thoughtco.com/introduction-to-green-technology-1991836
4. Green Technologies & Smart Cities. https://www.asteelflash.com/newsroom/green-technologies-smart-cities/
5. C. Yin, Z. Xiong, H. Chen, J. Wang, D. Cooper, B. David, A literature survey on smart cities. Science China Inf. Sci. **58**(10), 1–18 (2015)
6. R. Wenge, X. Zhang, C. Dave, L. Chao, S. Hao, Smart city architecture: A technology guide for implementation and design challenges. China Communications **11**(3), 56–69 (2014)
7. C. Harrison, B. Eckman, R. Hamilton, P. Hartswick, J. Kalagnanam, J. Paraszczak, P. Williams, Foundations for smarter cities. IBM J. Res. Dev. **54**(4), 1–16 (2010)
8. R. Giffinger, H. Gudrun, Smart cities ranking: An effective instrument for the positioning of the cities? ACE **4**(12), 7–26 (2010)
9. D. Washburn, U. Sindhu, S. Balaouras, R.A. Dines, N. Hayes, L.E. Nelson, Helping CIOs understand "smart city" initiatives. Growth **17**(2), 1–17 (2009)
10. S. Dirks, M. Keeling. *A Vision of Smarter Cities: How Cities Can Lead the Way Into a Prosperous and Sustainable Future*. (IBM Institute for business Value, 2009), p. 8
11. V. Javidroozi, H. Shah, A. Amini, A. Cole. *Smart City as an Integrated Enterprise: A Business Process Centric Framework Addressing Challenges in Systems Integration*. In Proceedings of 3rd International Conference on Smart Systems, Devices and Technologies, Paris (2014). pp. 55–59
12. R. Wantmure, M. Dhanawade. Use of Internet of Things for Building Smart Cities in India. In: *NCRD's Technical Review* (2016)
13. K.B. Ahmed, M. Bouhorma, M.B. Ahmed. *Age of Big Data and Smart Cities: Privacy Trade-Off*. arXiv preprint arXiv:1411.0087. (2014)
14. R. Kitchin, M. Dodge, The (in) security of smart cities: Vulnerabilities, risks, mitigation, and prevention. J. Urban Technol. **26**(2), 47–65 (2019)

15. A. Arroub, B. Zahi, E. Sabir, M. Sadik. *A Literature Review on Smart Cities: Paradigms, Opportunities and Open Problems*. In 2016 International Conference on Wireless Networks and Mobile Communications (WINCOM) (IEEE, 2016). pp. 180–186
16. H. Chourabi, T. Nam, S. Walker, J.R. Gil-Garcia, S. Mellouli, K. Nahon, H.J. Scholl. *Understanding Smart Cities: An Integrative Framework*. In: 2012 45th Hawaii international conference on system sciences (IEEE, 2012). pp. 2289–2297
17. T. Nam, T.A. Pardo. *Conceptualizing Smart City with Dimensions of Technology, People, and Institutions*. In Proceedings of the 12th annual international digital government research conference: digital government innovation in challenging times (ACM, 2011). pp. 282–291
18. Smart Cities Mission Is Still Very Much a Work in Progress Post Three Years Of Its Launch https://economictimes.indiatimes.com/news/economy/infrastructure/smart-cities-mission-is-still-very-much-a-work-in-progress-post-three-years-of-its-launch/articleshow/64523035.cms?from=mdr
19. F. Al-Turjman, H. Zahmatkesh, R. Shahroze. An overview of security and privacy in smart cities' IoT communications. Trans. Emerg. Telecommun. Technol., e3677 (2019)
20. G.T. Singh, F.M. Al-Turjman. *Cognitive Routing for Information-Centric Sensor Networks in Smart Cities*. In 2014 International Wireless Communications and Mobile Computing Conference (IWCMC) (IEEE, 2014). pp. 1124–1129)
21. U.D. Ulusar, D.G. Ozcan, F. Al-Turjman, Open source tools for machine learning with big data in smart cities, in *Smart Cities Performability, Cognition, & Security*, (Springer, Cham, 2020), pp. 153–168
22. M. Bhuvaneswari, G.N. Balaji, F. Al-Turjman, Machine learning parameter estimation in a smart-city paradigm for the medical field, in *Smart Cities Performability, Cognition, & Security*, (Springer, Cham, 2020), pp. 139–151
23. G. Başer, O. Doğan, F. Al-Turjman, Smart tourism destination in smart cities paradigm: A model for antalya, in *Artificial Intelligence in IoT*, (Springer, Cham, 2019), pp. 63–83
24. M. Sen, A. Dutt, J. Shah, S. Agarwal, A. Nath, Smart software and smart cities: A study on green software and green technology to develop a smart urbanized world. Int. J. Adv. Comp. Res. **2**(6), 373–380 (2012)
25. Mr. Harish Agarwal (ceo, Supreme & Co. Pvt. Ltd) Spoke on "understanding Smart Cities" At Rotary Club of Calcutta Mayfair on 22nd November2014 https://supremeco.wordpress.com/2014/12/06/mr-harish-agarwal-ceo-supreme-co-pvt-ltd-spoke-on-understanding-smart-cities-at-rotary-club-of-calcutta-mayfair-on-22nd-november2014/
26. Smart Switchgears EPR Magazine Editorial - https://www.eprmagazine.com/uncategorized/smart-switchgears/
27. Erp Magazine: Smart Power in Smart Cities https://www.nuos.in/news/erp-magazine-smart-power-in-smart-cities/
28. http://tec.gov.in/pdf/M2M/Design%20Planning%20Smart%20Cities%20with%20IoT%20ICT.pdf
29. Kec International Bags Orders Worth ₹1,323 Cr Our Bureau - https://www.thehindubusinessline.com/companies/kec-international-bags-orders-worth-1323-cr/article26508937.ece
30. J.S. Rao, M. Syamala, Internet of things (IoT) based Smart City architecture and its applications. Int. J. Comp. Math. Sci. **6**(10), 68–73 (2017)
31. A. Gaur, B. Scotney, G. Parr, S. McClean, Smart city architecture and its applications based on IoT. Proc. Comp. Sci. **52**, 1089–1094 (2015)

Chapter 7
Scheduling of Scientific Workflow in Distributed Cloud Environment Using Hybrid PSO Algorithm

Chetan Sharma and Mamoon Rashid

7.1 Introduction to Cloud Computing

Cloud computing is developing as the most recent conveyed computing worldview that gives a progressively versatile administration conveyance and utilization stage through virtualization of equipment and programming with the arrangement of devouring different administrations. It pulls in expanding interests of specialists in the region of Distributed and Parallel Computing and Service Oriented Computing [1]. Cloud computing conveys equipment framework and programming applications as administrations. It receives a market-arranged plan of action where clients are charged for devouring cloud administrations, for example, figuring, stockpiling and system administrations like ordinary utilities in regular day-to-day existence (e.g. water, power, gas and communication) on a pay-as-you-go premise. In the meantime, cloud service providers are committed to supplier attractive Quality of Service (QoS) in light of business benefit contracts [2]. There are challenges in cloud computing environments as well in the form of security, administration of resources, and its execution. Task scheduling in cloud systems is one such challenge in terms of resource management. Task scheduling aims to enhance execution of assignments and usage of resources in cloud systems [3].

Task scheduling is a standout among the most critical parts of cloud computing. Task is a bit of work to be executed in a predetermined time. Task scheduling is the way towards relegating assets to a specific task for the particular time for that assignment to be finished. The tasks are conveyed over resources in a suitable way with the end goal that essential inclinations between undertakings are met and add up to time expected to execute all tasks is limited. The primary point of task scheduling is to augment resource usage. It includes limiting sitting

C. Sharma · M. Rashid (✉)
School of Computer Science & Engineering, Lovely Professional University, Jalandhar, India

© Springer Nature Switzerland AG 2020
F. Al-Turjman (ed.), *Trends in Cloud-based IoT*, EAI/Springer Innovations
in Communication and Computing, https://doi.org/10.1007/978-3-030-40037-8_7

tight time for tasks. Task scheduling yields less task reaction time, so submitted tasks execution happens inside conceivable least time. Booking of tasks includes finding right succession for assignment execution under time limitations. Legitimate succession helps limiting aggregate time for errand execution. Legitimate task planning enhances effectiveness and execution of cloud condition. Keeping in mind the end goal to accomplish elite, different calculations for assignment booking have been proposed by specialists. The execution of cloud shifts with selection of the different calculation [4].

The outline of the chapter is structured as follows: Sect. 7.2 describes the related work on scheduling workflow in cloud environments using various algorithms. In Sect. 7.3, hybrid scheduling algorithm is proposed and explained. Section 7.4 discusses experimental setup and various results drawn. Conclusion and summary of research are given in Sect. 7.5.

7.2   Background

Scheduling of workflow in distributed cloud environments is done in number of works by using different optimization techniques in state of the art. A heuristic approach for scheduling application on cloud systems is presented by using particle swarm optimization (PSO). This research claims better computation in costs and savings in terms of communication cost [5]. To minimize the processing cost of task scheduling, an attempt is made by proposing a model with the use of particle swarm optimization (PSO) where rule based on small positions is used. This research claims better results for task scheduling when applied on cloud systems [6]. Workflow scheduling is further refined in cloud systems by using shuffled frog leaping algorithm. This algorithm provided better results in minimizing execution costs [7].

Work scheduling in cloud systems is optimized by using cat swarm optimization in [8]. The proposed technique in this research addresses both executions costs and transmission cost of data. The cost of tasks depends on balancing by reducing the makespan time in cloud systems using hybrid algorithm based on genetic and particle swarm optimization. This algorithm achieved optimal results compared to algorithms in state of the art [9]. Another attempt is made to improve scheduling of workflow in cloud systems with the help of BAT algorithm. This research achieved results with minimal overall workflow costs [10]. For the execution of programs in cloud systems, reliable and cost-efficient scheduling for tasks method is proposed. The algorithm used in this work proved to be efficient in terms of reducing costs [11]. Simulation model is implemented for efficient scheduling workflow in cloud systems. In this research scheduling algorithm is applied for implemented model which performed better in terms of scheduling with respect to state of the art [12]. Estimation of execution time for workflows is presented in terms of novel prediction model in [13]. This research claims that error in terms of execution time resulted from proposed prediction model is below 20%. For scheduling tasks on cloud systems, an efficient algorithm is proposed which minimizes the cost and makespan time [14]. Scheduling workflow in cloud systems is improved by proposing novel fault tolerant approach where emphasis is given on replication technique of tasks in cloud systems [15]. Another heuristic scheduling algorithm is presented for solving the scheduling in cloud systems efficiently. This research implemented this algorithm on Hadoop and cloud Sim and successfully reduced the makespan time in scheduling tasks on cloud systems [16]. Fuzzy-based clustering is used for scheduling the workflow of tasks in cloud computing environments. This research achieved the objective of minimizing makespan and scheduling time [17]. Algorithm for the estimation of workflow execution is proposed where problems of task scheduling are addressed. This research provided a solution for workflow of tasks with major emphasis on error management [18]. A solution for minimizing turnaround time is proposed within budget for workflow in cloud systems with two novel algorithms. The first scheduling algorithm used in this research provided better performance when the time budget is low and second scheduling algorithm proves to be better when cost of budget was high [19]. The resting cost of cloud workflows is minimized in cloud systems by proposing bag-based delay method. This method addressed the execution times in task scheduling [20].

An attempt is made to modify Heterogeneous Earliest Finish Time (HEFT) by reducing the makespan time of applications and distributing the workload effectively among processors. This chapter claims the reduction in makespan time and load balancing issues in comparison to HEFT and CPOP algorithms [21]. Detailed survey and analysis of workflow scheduling schemes is given in [22]. This chapter provides a classification of proposed scheme on the basis of use of scheduling algorithm for each scheme. Moreover, authors in this chapter have given comparison for all these schemes. Operations on a workflow graph are refined for increasing the performance of the algorithms in [18]. The authors in this chapter integrated their schema with scheduling algorithm GAHEFT. A novel algorithm is proposed for optimizing the scheduling of workflows in cloud [23]. The authors have tested this algorithm for different simulated cloud environments with the inclusion of different cost models. A hybrid algorithm based on genetic algorithm and heuristic methods is proposed in [24]. The authors proposed Likewise Earliest Finish Time as a substitute for HEFT. The mathematical model is provided for scheduling problems in work-flow environments [25]. The authors have compared model with existing PSO algorithm and proved that task allocation is done in optimal way. BAT algorithm is used to address the scheduling problems in clouds by maximizing the reliability and minimizing the execution time [26]. The authors in this chapter addressed these issues by keeping budget with the limits of user. Security model for data storage on third-party cloud server for managing workflow is proposed in [27]. This research showed how unknown users are denied the access for cloud systems using their proposed model taking care of constraints. A method to address the security issues in cloud environment is again provided in [28]. The author in this chapter has used software engineering principles to provide security for various cloud stored applications. The efficient storage of data in cloud systems is proposed in [29]. The authors in this research have used techniques based on

IP and geographic locations for the efficient access of stored data. An efficient algorithm is presented for task scheduling which addresses the issues of allocation of resources in cloud-based IoT systems [30]. The authors made use of fully informed particle swarm optimization and canonical particle swarm optimization. This work improved the quality of service in scheduling by considering delay and throughput. An optimized algorithm is proposed for improving the Quality of Service in Internet of Things environment [31]. This research addresses the fault tolerance in routing and has verified algorithm for its efficiency. Performance issues related to implementing of IoT-based applications has been presented in [32]. This work has summarized the challenges which are required to be considered for improvement in performance of IoT systems.

7.3 Hybrid PSO Algorithm

Task scheduling is one of the main challenge in cloud workflows. Researchers around the globe have proposed and implemented number of heuristic algorithms which are good enough in solving the problems of task scheduling in cloud systems. Every research tries to refine these algorithms in two possible ways. One in terms of Quality of Service workflow scheduling and another best effort workflow scheduling. In this research, the authors proposed and implemented hybrid algorithm where population is generated for Particle Swarm Optimization (PSO) using the Predict Earliest Finish Time algorithm (PEFT). The authors used makespan time and cost as parameters to evaluate the results of the proposed approach. The various steps in proposed hybrid algorithm is shown in Fig. 7.1. All steps of this algorithm are given in terms of steps in this section.

The major challenge of scheduling is mapping of scientific tasks to available resources in cloud to minimize cost and execution time keeping deadline as a constraint. The authors in this research tried to focus on model generation which consumes less cost and execution time than the other existing approaches of scheduling.

1. Generate an initial population using the PEFT algorithm.

   (a) Compute the Optimistic Cost Table (OCT). OCT is represented in matrix form where tasks act as rows of a matrix and virtual machines as columns of matrix. The value of OCT is calculated with backward approach recursively according to Eq. (7.1). It will compute the execution cost of all children tasks of current task till it will reach the last point.

$$
\begin{aligned}
\text{OCT}\,(t_i, p_k) &= \max \quad t_j \in \text{succ}\,( ) \\
&\left[\min_{\ p_w\, \in\ P}\ \left\{ \text{OCT}\,\left(t_j,\ p_w\right) + \left(t_j,\ p_w\right) + \left(\overline{C_{i,j}}\right) \right\} \right] \\
&\text{and } \overline{C_{i,j}} = 0, \quad \text{if } p_w = p_k
\end{aligned}
\tag{7.1}
$$

where
$\overline{C_{i,j}}$ = average communication cost

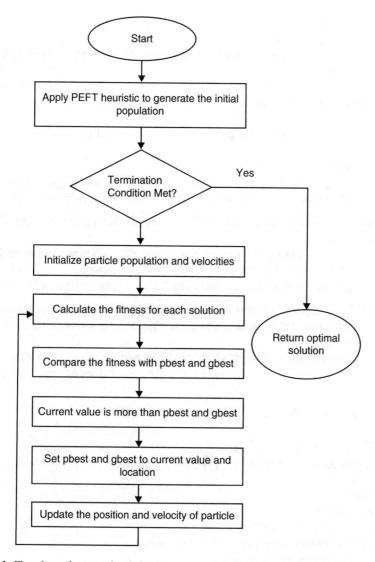

Fig. 7.1 Flowchart of proposed technique

t_i, t_j = different tasks running

P_w, p_k = different processors active during the process.

(b) For every node the cumulative Optimistic Cost Table is computed and it defines the rank of every node or task (rank_{oct}) as per Eq. (7.2). The order of tasks in the list comes in the decreasing order of rank_{oct}.

$$\text{rank}_{oct}(t_i) = \frac{\sum_{k=1}^{P} \text{OCT}(t_i, p_k)}{P} \tag{7.2}$$

where:

$\text{rank}_{\text{oct}}(t_i)$ = Average OCT of each task.

(c) Repeat steps (a) and (b) till list is un-empty, else best schedule of makespan is returned.

(d) Calculate Optimistic Earliest Finish Time is using Eq. (7.3) for allocating a task to processor on the basis of insertion-based policy.

$$O_{\text{EFT}}(t_i, p_j) = \text{EFT}(t_i, p_j) + \text{OCT}(t_i, p_j) \tag{7.3}$$

where:

O_{EFT} = Optimistic Earliest Finish Time ith task and jth processor. Processor is assigned with task which provides the minimum O_{EFT}.

2. Return to optimal solution if the condition of termination is reached, else steps 3 to 6 are repeated.
3. Generated high-quality schedule is seeded into the PSO algorithm as an initial particle population. Velocity of particles is initialled as well.
4. Fitness of each particle is calculated using the makespan time of the schedule.

   (a) Generated particle fitness is compared with particle's pbest, and if its current value is better in comparison to pbest, then pbest is set as current value and location.
   (b) Generated particle fitness is compared with particle's gbest, and if its current value is better in comparison to gbest, then gbest is set as current value and location.

5. Update the velocity and position of particle using Eq. (7.4)

$$\begin{aligned} \vec{x_l}(t+1) &= \vec{x_l}(t) + \vec{v_l}(t) \\ \vec{v_l}(t+1) &= \omega \cdot \vec{v_l}(t) + c_1 r_1 \left(\vec{y_l}(t) - \vec{x_l}(t) \right) + c_2 r_2 \left(y(t) - \vec{x_l}(t) \right) \end{aligned} \tag{7.4}$$

where:

ω = inertia;

C_i = acceleration coefficient; $i = 1; 2$

r_i = random number; $i = 1; 2$ and $r_i\ 2 = 0; 1$

$\vec{y_i}$ = best position of particle i

$\vec{y}$ = position of the best particle in the population

$\vec{x_i}$ = current position of particle i

6. Repeat step 4 until the stopping criteria is met.

7.4 Experimental Results

This section presents experimental results of the proposed technique. The proposed work is implemented using netbeans workflow simulation. Two parameters, makespan time and cost, are used to evaluate the performance of the proposed technique. Makespan time is the total processing time of all the tasks and cost is the cost required to process all the tasks. The inputs are supplied to the implemented PPSO algorithm in the form of Montage_100 and Heterogeneous Earliest Finish Time (HEFT) xml files.

Table 7.1 shows makespan time results achieved with Predict Earliest Finish Time based PSO (PPSO) and results achieved with existing technique of PSO on the basis of deadline factor of 100 scenarios. Alpha is the inertia factor (Weight), which plays crucial role in deciding or calculating that how much previous values will impact the current values as described in PSO of start of the art.

Figure 7.2 shows the comparison of makespan time for Predict Earliest Finish Time based PSO (PPSO) with PSO on the basis of deadline factor of 100 scenarios. It is clear from Fig. 7.2 that the PPSO takes less time than the PSO.

Table 7.2 shows cost evaluation results achieved with Predict Earliest Finish Time based PSO (PPSO) and results achieved with existing technique of PSO on the basis of deadline factor of 100 scenarios.

Table 7.1 Makespan time results of PSO and PPSO for 100 scenarios on the basis of deadline factor

| | Montage_100 (time) | |
Inertia factor(weight) alpha	Particle swarm optimization PSO	Predict earliest finish time based PSO (PPSO)
2	318.38	279.49
2.5	406.3	304.82
3	377.21	317.46
3.5	314.92	279.05
4	342.39	281.47
4.5	368.01	267.03

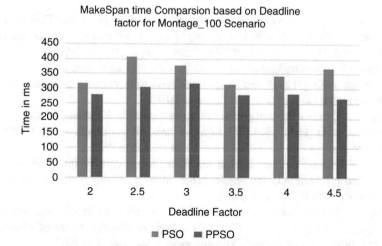

Fig. 7.2 Makespan time comparison of 100 scenarios on the basis of deadline factor

Table 7.2 Cost evaluation results of PPSO and PSO for 100 scenarios on the basis of deadline factor

Alpha	Montage_100 (cost)	
	PSO	PPSO
2	1439.6	1246.67
2.5	1417.25	1240.9
3	1444.02	1215.67
3.5	1440.7	1249.48
4	1446.1	1255.69
4.5	368.01	267.03

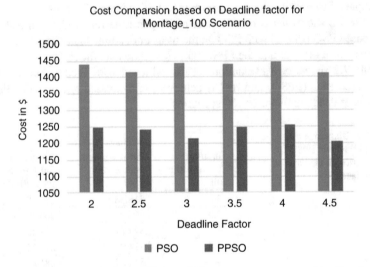

Fig. 7.3 Cost comparison of PPSO and PSO for 100 scenarios on the basis of deadline factor

Figure 7.3 shows the comparison of cost evaluation for Predict Earliest Finish Time based PSO (PPSO) with PSO on the basis of deadline factor of 100 scenarios. It is clear from Fig. 7.3 that the PPSO processes with less cost than PSO.

Table 7.3 shows makespan time results achieved with Predict Earliest Finish Time based PSO (PPSO) and results achieved with existing technique of PSO technique on the basis of deadline factor of HEFT.

Figure 7.4 shows the comparison of makespan time for Predict Earliest Finish Time based PSO (PPSO) with PSO on the basis of deadline factor of HEFT. It is clear from Fig. 7.4 that the PPSO require less time than PSO.

Table 7.4 shows cost evaluation results achieved with Predict Earliest Finish Time based PSO (PPSO) and results achieved with existing technique of PSO on the basis of deadline factor of HEFT.

Figure 7.5 shows the comparison of cost evaluation for Predict Earliest Finish Time based PSO (PPSO) with PSO on the basis of deadline factor of HEFT. It is clear from Fig. 7.5 that the PPSO processes with less cost than PSO.

	HEFT (time)	
Alpha	PSO	PPSO
2	81.76	62.09
2.5	86.95	71.91
3	99.57	70.34
3.5	75.42	61.44
4	76.08	63.18
4.5	75.21	74.32

Table 7.3 Makespan time results of PPSO and PSO for HEFT on the basis of deadline factor

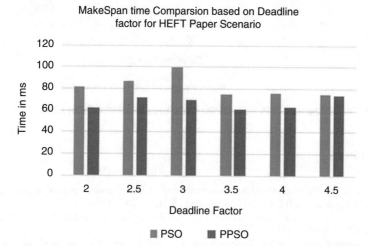

Fig. 7.4 Makespan time comparison of PPSO and PSO for HEFT on the basis of deadline factor

Table 7.4 Cost evaluation of PPSO and PSO on the basis of deadline factor HEFT

	HEFT (cost)	
Alpha	PSO	PPSO
2	93.2	66.25
2.5	116.76	89.53
3	93.03	61.83
3.5	77.45	72.15
4	90.49	62.52
4.5	79.92	79.47

7.5 Conclusion

Task scheduling plays significant role in improving the performance of cloud computing. This research work generates population for PSO using the PEFT Predict Earliest Finish Time algorithm. This work aimed to obtain an optimized schedule by PSO using PEFT population which reduced the cost and execution time. Two parameters, makespan time and cost, were used to evaluate the performance of the proposed technique. Experimental results show that the proposed technique require less time and cost than the existing technique in state of the art and outperformed.

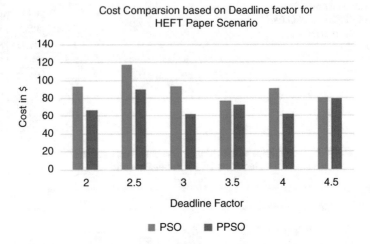

Fig. 7.5 Cost comparison of PPSO and PSO on the basis of deadline factor HEFT

References

1. A. Verma, S. Kaushal, Deadline constraint heuristic-based genetic algorithm for workflow scheduling in cloud. Int. J. Grid Utility Comp. **5**(2), 96–106 (2014)
2. A. Kaur, A review of workflow scheduling in cloud computing environment. Int. J. Comp. Sci. Eng. **4**(2), 47–50 (2015)
3. S.A. Hamad, F.A. Omara, Genetic-based task scheduling algorithm in cloud computing environment. Int. J. Adv. Comput. Sci. Appl. **7**(4), 550–556 (2016)
4. A.E. Keshk, Cloud computing online scheduling. Egypt. IOSR J. Eng. **4**(03), 7–17 (2014)
5. S. Pandey, L. Wu, S.M. Guru, R. Buyya. *A Particle Swarm Optimization-Based Heuristic for Scheduling Workflow Applications in Cloud Computing Environments*. In 2010 24th IEEE international conference on advanced information networking and applications (IEEE, 2010). pp. 400–407
6. L. Guo, S. Zhao, S. Shen, C. Jiang, Task scheduling optimization in cloud computing based on heuristic algorithm. J. Networks **7**(3), 547 (2012)
7. P. Kaur, S. Mehta, Resource provisioning and work flow scheduling in clouds using augmented shuffled frog leaping algorithm. J. Paral Distr Comp **101**, 41–50 (2017)
8. S. Bilgaiyan, S. Sagnika, M. Das. *Workflow Scheduling in Cloud Computing Environment Using Cat Swarm Optimization*. In 2014 IEEE International Advance Computing Conference (IACC) (IEEE, 2014). pp. 680–685
9. A.M. Manasrah, H. Ba Ali, Workflow scheduling using hybrid ga-pso algorithm in cloud computing. Wirel. Commun. Mob. Comput. **2018** (2018)
10. S. Raghavan, P. Sarwesh, C. Marimuthu, K. Chandrasekaran. *Bat Algorithm for Scheduling Workflow Applications in Cloud*. In 2015 International Conference on Electronic Design, Computer Networks & Automated Verification (EDCAV) (IEEE, 2015). pp. 139–144
11. S. Su, J. Li, Q. Huang, X. Huang, K. Shuang, J. Wang, Cost-efficient task scheduling for executing large programs in the cloud. Parallel Comput. **39**(4–5), 177–188 (2013)
12. P. Bryk, M. Malawski, G. Juve, E. Deelman, Storage-aware algorithms for scheduling of workflow ensembles in clouds. J. Grid Comp. **14**(2), 359–378 (2016)
13. I. Pietri, G. Juve, E. Deelman, R. Sakellariou. *A Performance Model to Estimate Execution Time of Scientific Workflows on the Cloud*. In 2014 9th Workshop on Workflows in Support of

Large-Scale Science (IEEE, 2014). pp. 11–19

14. D. Poola, S.K. Garg, R. Buyya, Y. Yang, K. Ramamohanarao. *Robust Scheduling of Scientific Workflows with Deadline and Budget Constraints in Clouds*. In 2014 IEEE 28th international conference on advanced information networking and applications (IEEE, 2014). pp. 858–865

15. K. Ganga, S. Karthik. *A Fault Tolerent Approach in Scientific Workflow Systems Based on Cloud Computing*. In 2013 International Conference on Pattern Recognition, Informatics and Mobile Engineering (IEEE, 2013). pp. 387–390

16. C.W. Tsai, W.C. Huang, M.H. Chiang, M.C. Chiang, C.S. Yang, A hyper-heuristic scheduling algorithm for cloud. IEEE Trans. Cloud Comp. **2**(2), 236–250 (2014)

17. F. Guo, L. Yu, S. Tian, J. Yu, A workflow task scheduling algorithm based on the resources' fuzzy clustering in cloud computing environment. Int. J. Commun. Syst. **28**(6), 1053–1067 (2015)

18. A.M. Chirkin, A.S. Belloum, S.V. Kovalchuk, M.X. Makkes, M.A. Melnik, A.A. Visheratin, D.A. Nasonov, Execution time estimation for workflow scheduling. Futur. Gener. Comput. Syst. **75**, 376–387 (2017)

19. M. Mao, M. Humphrey. *Scaling and Scheduling to Maximize Application Performance Within Budget Constraints in Cloud Workflows*. In 2013 IEEE 27th International Symposium on Parallel and Distributed Processing (IEEE, 2013). pp. 67–78

20. Z. Cai, X. Li, R. Ruiz, Q. Li, A delay-based dynamic scheduling algorithm for bag-of-task workflows with stochastic task execution times in clouds. Futur. Gener. Comput. Syst. **71**, 57–72 (2017)

21. K. Dubey, M. Kumar, S.C. Sharma, Modified HEFT algorithm for task scheduling in cloud environment. Proc. Comp. Sci. **125**, 725–732 (2018)

22. M. Masdari, S. ValiKardan, Z. Shahi, S.I. Azar, Towards workflow scheduling in cloud computing: A comprehensive analysis. J. Netw. Comput. Appl. **66**, 64–82 (2016)

23. S. Elsherbiny, E. Eldaydamony, M. Alrahmawy, A.E. Reyad, An extended intelligent water drops algorithm for workflow scheduling in cloud computing environment. Egypt. Inform. J. **19**(1), 33–55 (2018)

24. D. Nasonov, N. Butakov, M. Balakhontseva, K. Knyazkov, A.V. Boukhanovsky. *Hybrid Evolutionary Workflow Scheduling Algorithm for Dynamic Heterogeneous Distributed Computational Environment*. In International Joint Conference SOCO'14-CISIS'14-ICEUTE'14 (Springer, Cham, 2014). pp. 83–92

25. M. Kumar, S.C. Sharma. PSO-COGENT: Cost and Energy Efficient scheduling in Cloud environment with deadline constraint. In *Sustainable Computing: Informatics and Systems* (2018)

26. N. Kaur, S. Singh, A budget-constrained time and reliability optimization BAT algorithm for scheduling workflow applications in clouds. Proc. Comp. Sci. **98**, 199–204 (2016)

27. M. Rashid, E.R. Chawla, Securing data storage by extending role-based access control. Int. J. Cloud Appl. Comp. **3**(4), 28–37 (2013)

28. S. Aljawarneh, Cloud security engineering: Avoiding security threats the right way, in *Cloud Computing Advancements in Design, Implementation, and Technologies*, (IGI Global, Pennsylvania, 2013), pp. 147–153

29. M. Rashid, H. Singh, V. Goyal, Cloud storage privacy in health care systems based on IP and geo-location validation using K-mean clustering technique. Int. J. E-Health Med. Commun. **10**(4), 54–65 (2019)

30. F. Al-Turjman, M.Z. Hasan, H. Al-Rizzo, Task scheduling in cloud-based survivability applications using swarm optimization in IoT. Trans. Emerg. Telecommun. Technol. **30**(8), e3539 (2019)

31. M.Z. Hasan, F. Al-Turjman, SWARM-based data delivery in social internet of things. Futur. Gener. Comput. Syst. **92**, 821–836 (2019)

32. F. Al-Turjman, H. Zahmatkesh, R. Shahroze, An overview of security and privacy in smart cities' IoT communications. Trans. Emerg. Telecommun. Technol., e3677 (2019)

Chapter 8
Model-Based Recommender Systems

Bahrudin Hrnjica, Denis Music, and Selver Softic

8.1 Introduction

We are facing information growth every day, and handling that information is not an easy task. Information overflow is present in every segment of human activities, which makes it difficult for customers to decide which information is good and relevant for them. In order to write this chapter, the authors tried to find relevant information through the internet. The internet search engine google.com returns nearly 300 million results for "recommender systems," or 42 million results for "content-based filtering." Is it necessary for the authors to check all the results before writing this chapter? Obviously not, because to do so it requires several hundred years. A similar scenario can be described when the authors try to buy a relevant book in order to write this article. They visited amazon.com and tried to find books about recommender systems. When the term "recommender systems" was entered, the Amazon search engine returned more than 300 books. Should all books the authors must read in order to write this chapter? Obviously not, because they need at least a year to read those books. Those are only a few examples where the information overflow is described, and it may make difficulties for the user in selecting relevant information. The same information overflow issue will come out

B. Hrnjica (✉)
University of Bihac, Bihac, Bosnia and Herzegovina
e-mail: bahrudin.hrnjica@unbi.ba

D. Music
Faculty of Information Technology, University "Dzemal Bijedic" Mostar, Mostar, Bosnia and Herzegovina
e-mail: denis.music@fit.ba

S. Softic
IT and Business Informatics, CAMPUS 02 University of Applied Sciences, Graz, Austria
e-mail: selver.softic@campus02.at

© Springer Nature Switzerland AG 2020
F. Al-Turjman (ed.), *Trends in Cloud-based IoT*, EAI/Springer Innovations
in Communication and Computing, https://doi.org/10.1007/978-3-030-40037-8_8

in other scenarios, e.g., supermarket, restaurants, cinema, retail stores, etc. So, the natural question can be stated: how to overcome the information overflow when trying to find or more likely purchase specific information or item? Is there any system which can recommend few instead of returning all information? How the user is sure the item which is trying to buy is relevant and satisfied user's needs. Is it possible that the information can be filtered in order to be reduced and provided in much less amount, so the user can evaluate it and select the appropriate ones? The answer to the questions is recommender systems. The recommender systems are nothing but the systems for reducing the information overflow and providing recommendation instead of search result. They are algorithms for information filtering based on relevance estimation.

8.1.1 Brief History of the Recommender Systems

First works of the recommender systems appeared in the early 1990s when the internet and email started to be massively used and popularized. One of the first works about recommender systems was trying to reduce the amount of incoming documents due to the increasing use of electronic mail [1]. The systems were able to filter the content based on the users' activities and collaborations. In the middle of 1990s the first distributed recommended systems published on the network. For the first time the GroupLens company performed automated collaborative filtering for the internet site Usenet [2]. At the same time, the RINGO systems implemented music albums and artists recommender [3], while the Bellcore video recommender implemented the first online movie recommender [4]. In the beginning the used algorithm was based on users' inputs where similar users' behaviors were classified by using "k-nearest neighbor" algorithms. At the end of the twentieth century companies started online store and they became very popular. Millions of users wanted to buy products using internet. However, the number of products was exponentially increased, and for the users it was very hard to find suitable products. One of the first companies that used the recommender systems in online store was Amazon [5]. In 1998 Amazon launched recommender system enabling personalized recommendation catalog to millions of users. Little later other companies, e.g., YouTube [6], Netflix [7], and many others started adopting the algorithm for their products and customers. Unlike previously well-popularized user-based recommender system, the algorithm was based on products not users. The algorithm searched related items for each item in the catalog. The algorithm can be simply explained as: *people who buy one item are unusually likely to buy the other* [8]. In the first several years of the twenty-first century research in recommender systems was intensified specially after Netflix announced the one million dollar prize for a 10% improvement in prediction accuracy of the current recommended system [9]. That was a milestone in the development of the recommender systems, since the competition succeeded in bringing many developers and engineers together to work on improvements of their recommended system. In this time Facebook and then little later Twitter became the main social stream, and everybody wanted to be on Facebook and Twitter and find their friends and followers. So, the recommended systems were improved in order to recommend similar friends by the same location, education, and other properties. Both companies started to develop their recommender systems. One of the first recommender was Group Recommendation Systems for Facebook, where the algorithm was able to recommend group using a combination of hierarchical clustering technique and decision tree [10]. The Twitter company developed new recommender systems based on the user preferences [11], right after it has become one of the two most popular social networks. Recent advancement in deep learning pushed once again recommender systems in a new track, and engineers have started building recommender systems using deep learning. There are several review papers in which the reader may find the latest advancement of the deep learning in recommender systems [12, 13]. Using deep learning technique in recommender systems opened a wide variety of applications with specific approaches and purposes [14–16].

Model based recommender systems have application on devices which do not require full computing power, since the model has already trained. This brings a new direction in application development based on cloud computing and Internet of Things, IoT. Putting intelligence in form of ML models on IoT devices provide application development to play a main role in smart cities. In order to protect the network by delivering the information safety and accurately IoT devices are facing many challenges in term of QoS, resource management but most important in security [38, 39]. Despite security challenges, IoT applications have found application in a variety of intelligent solutions especially in smart cities. [40, 41].

8.1.2 Recommender Systems

The basic idea of the recommended systems is based on the assumption that if the users share the same behaviors while buying in the online stores, selecting the similar artists, reading the same news, or watching the same movies, they will also have similar behaviors in the future. In other words, the recommender systems estimate users' preferences on items and recommended items, so that other users probably may like them. The recommended systems can analyze items in two ways: they analyze items in terms of how users like them, so that similar users will like same products, or analyze items in terms of how items are popularized between users, regardless of the users' similarities. Based on the two fundamental concepts, the recommender systems can be classified into three main types: content-based, collaborative filtering, and hybrid-based systems [17, 18] (Fig. 8.1).

Content-based recommender systems analyze set of descriptions of items previously rated by users and create a profile of a user behavior based on the attributes of the items rated by that user. For example, a user rates a book at an online book store. The book has its attributes, e.g., genre, title, author, type, prize. The attributes are used by the recommender systems in a way that it can provide the recommendation list for the user.

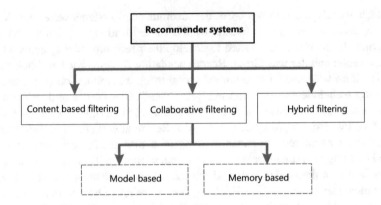

Fig. 8.1 Categorization of recommender systems

Collaborative filtering based recommenders analyze users activities and try to find similarities between users, how they like, view, or buy the same items. In that case users who share the same behavior in the past will probably have similar tastes in the future. For example, if the user 1 and user 2 have very similar purchase history, e.g., they liked items from the same category, bought the same items, and user 1 has recently bought a movie that user 2 has not yet watched. Naturally, it would be good to recommend the movie to user 2.

Hybrid based recommender systems combine previous two techniques to generate better and more precise recommendations. For example, there is information about users purchasing history with details about individual items, and users. A recommender system could be enhanced by hybridizing collaborative or social filtering with content-based techniques, in order to overcome the problems of the previous types.

The formulation of the recommendation can be defined as *predicting the rating value* for a user-item combination or *prediction of the top k items* for a particular user. The first formulation is more general and requires the user preferences over the set of items. For M users and N items the training data is provided as sparse matrix. The missing values are subject to prediction. The problem is also referred to as the *matrix completion problem*.

The recommender starts by defining a sparse rating matrix R with M users and N items. The goal is to create predicted dense interaction matrix $\hat{R}$ with all ratings of users and items. Let r_{u_i} denote the preference of a user u to item i so that r_{u_i} represents the prediction score. Since the sparse rating matrix R is not complete, two vectors are created.

$$\mathbf{r^u} = r^{u_1}, r^{u_2}, \dots, r^{u_{N-1}}, r^{u_N}, \tag{8.1}$$

and the second vector (columns of R) represents partially observed rating $\mathbf{r^i}$ to represent each item:

$$\mathbf{r^i} = \begin{bmatrix} r^{i_1} \\ r^{i_2} \\ \dots \\ r^{i_M} \end{bmatrix}. \tag{8.2}$$

Since the matrix is sparse two sets are created: O and O^- denote the observed and unobserved interaction sets. Let k denote latent factor that represents latent dimension so that for two matrices $U \in \Re^{M \times k}$ and $V \in \Re^{k \times N}$ can be defined so that they satisfied:

$$\underbrace{\mathbf{R}}_{M \times N} = \underbrace{\mathbf{U}}_{M \times k} \times \underbrace{\mathbf{V}}_{k \times N}. \tag{8.3}$$

The prediction of the top k items put less general into more specific problem. The problem can be calculated by calculating the first formulation and then rank the prediction of the top k-items. The primary goal of recommender systems is increasing the sales and is utilized by merchants to increase their profit. The goal can be achieved by bringing relevant items to the attention of users, so that it increases the volume sales and profits for the merchant. On the other hand, recommendation can help users to improve the overall satisfaction of the web site. If the user receives relevant recommendation each time he visits the web site, then he will be more satisfied with it and will use it again. In this way the user will develop loyalty with the web site, but also for the merchants indicators of users' needs, so the recommendation can be further improved. Not only sale is the recommender systems used for. In case of Facebook or Twitter recommendation is based on social connections, which have indirect benefits to the site by increasing its advertising profits.

8.1.3   Content-Based Recommender Systems

Content-based filtering, (CBF) recommender systems allow users to filter information based on a set of methods that perform analysis in order to predict what the user might be interested in, or what is relevant to the user. Many of the existing sites generate a recommendation based on the content of the product. In such cases algorithms allow site owner or administrator to determine rules, often based on user demographics or other personal characteristics. The rules are used to influence the content served to the user whose profile complies with one or more rules. Continuing with the book example, CBF should recommend *Book2* to users who liked *Book1* only if the most significant characteristics (attributes) of those books are similar. Therefore, it is very important to identify which book characteristics to observe, because some of them are very significant as they affect user preferences (content, genre, writer, publisher, etc.), while others can mostly be ignored (ISBN, how many

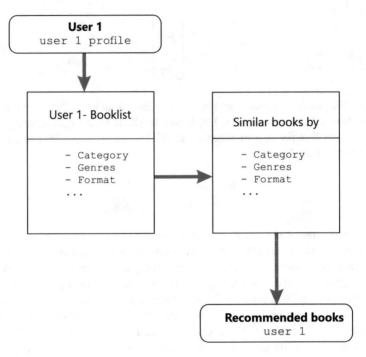

Fig. 8.2 Concept of content-based recommender system

years author spend in writing a book, number of pages, etc.). After selecting the most significant ones, it is necessary to determine the ways to calculate similarity between attributes which would ultimately represent a similarity between the items (books). In one of the simplest scenarios, attributes would have ratings which could be used as coordinate values to position item in the n-dimensional space (if only three attribute values are considered, then the 3D space would be observed) (Fig. 8.2).

Simplified examples of how CBF systems work might include analysis of data about five books ($b_1, b_2, \ldots, b_5$), with three attributes being evaluated for each book (x_1, x_2, x_3). Figure 8.3 shows the ratings of each of the attribute according to which books are positioned in 3D space. The most similar books should be those between which there is the shortest distance (Table 8.1).

The distance between two points is the length of the straight line connecting them. Since in the previous example, only three attributes were observed, its representation of distances in 3D space is very simple. However, it is quite difficult to imagine an example with more attributes that would be represented in multidimensional space. In order to introduce more complex types of interpoint relationships (for m points which represent items or books), the distances calculated between all possible pairs is stored inside the distance matrix of size $m \times m$.

Fig. 8.3 Determining similarities between items in 3D space

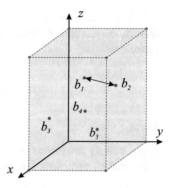

Table 8.1 Determining similarities between items in 3D space

	x_3	x_3	x_3
b_1	5	3	4
b_2	3	3	4
b_3	1	1	4
b_4	3	3	1
b_5	1	2	5

In order to determine the distance (which represents similarity) between the books, several metrics are available: Euclidean, Hamming, and correlation distance. Which of the aforementioned metrics will be used depends primarily on the context and nature of data. In order to find similarity between two observed items or books, attribute values are represented by vectors X and Y.

$$X = (x_1, x_2, \ldots, x_n) \qquad Y = (y_1, y_2, \ldots, y_n). \tag{8.4}$$

Euclidean distance is only appropriate for data measured on the same scale and it can be described as the length of the line segment that connects two coordinates. Book attribute values from the previous example were interpret as coordinates in 3D space and therefore Euclidean distance was the simplest metric to calculate similarity. Euclidean distance can be represented as the sum of squared differences of two vectors:

$$d_E(X, Y) = \sqrt{\sum_i^n (x_i - y_i)^2}. \tag{8.5}$$

Hamming distance results with positive integer that represents the number of changes required to convert one data point into another, or as the number of positions where the symbols x_i and y_i differ:

$$d_H(X, Y) = \sum_{i=1}^n \delta(x_i, y_i); \qquad \delta(x_i, y_i) = \begin{cases} 0 & x_i = y_i \\ 1 & x_i \neq y_i \end{cases}. \tag{8.6}$$

While Euclidean distance represents the sum of squared differences of two vectors, correlation distance can be represented as the average product of the two vectors:

$$d_C(X, Y) = \frac{\frac{1}{n} \sum_i x_i y_i - \mu_X \mu_Y}{\sigma_X \sigma_Y}, \qquad (8.7)$$

where,

- μ_X and μ_Y are means of vectors X and Y, respectively,
- σ_X and σ_Y represent the standard deviations of X and Y.

The numerator in the expression (8.7) is called the covariance of X and Y, and is the difference between the mean of the product of X and Y subtracted from the product of the means. If vectors X and Y are standardized, they will each have a mean of 0 and a standard deviation of 1, and therefore formula can be significantly reduced.

Cold Start Problem of the CBF

As discussed in [19] CBF recommender systems have several key shortcomings that need to be addressed. Because of the way the system works, it is necessary for the user to evaluate enough content to allow the system to make appropriate recommendations which is known as the cold start problem. Also, the disadvantage of such systems is that the recommended content is often not diverse. Because of the way they determine the usefulness of a product, where the characteristics of an individual are compared products with a customer problem, and only recommendations for content for which this value is maximal or very high, it happens that products that differ in content from what the user has reviewed previously are never recommended to the user. This drawback is also called content overspecialization or serendipity problem to highlight the tendency of the CBF systems to produce recommendations with a limited degree of novelty. To give an example, when a user has only rated book written by Jamie McGuire, system will recommend mostly that kind of books and rarely finding something novel. The drawback is directly connected with the fact that enough ratings have to be collected before a CBF recommender system can really understand user preferences and provide accurate recommendations. Therefore, when few ratings are available, as for a new user, the system will not be able to provide reliable recommendations [42]. In other words, new items or books without any ratings cannot be recommended while new users who did not share their preferences with the recommended system yet cannot expect any excessively personalized recommendations. Some solutions which eliminate these shortcomings to a greater or lesser extent have been proposed [19], all of which depend on the context and type of data used.

8.1.4   Collaborative Filtering Recommender Systems

Collaborative filtering (CF) recommender systems aim to predict the usefulness of an individual item for a user based on the evaluation of that item by other system users. With the CF, the ratings that users have assigned to an item are used as an approximate presentation of their interests and needs. Unlike CBF recommendations, CF user model does not contain item rated data but rather the grades assigned by the target user are compared with the ratings that have been assigned by other users of the system. This creates the assumptions for determining the closest set of items that may be of interest to the user. Thus, in addition to the target user model, a database with models of other system users is very important. System will recommend items which are not rated by the current user but are highly rated by similar users. In other words, CF is based on the expectation that people who had similar interests in the past are likely to have similar interests again in the future (Fig. 8.4).

In general, one can distinguish memory-based and model-based technique.

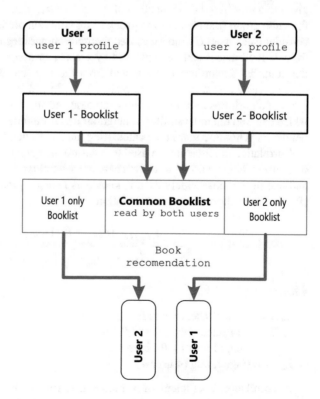

Fig. 8.4 Concept of collaborative filtering recommender systems

Memory-Based Filtering Technique

Memory-based approach loads the entire database into system memory and makes prediction based upon the in-line memory database (see [20]). The main disadvantage of the memory-based approach is the requirement of loading a large amount of in-line memory. This causes problems especially when rating matrix becomes so huge in situations where there are extremely many persons using a system. High consumption of computational lowers the system performance and decreases the responsiveness towards the user requests.

Memory-based filtering technique can be classified through one of the two basic approaches called:

– User-based CF and
– Item-based CF.

User-based CF, also called memory-based, primarily focuses on the system users and they perform the main role. If majority of the users have the same interests, then they are eligible to join into one group $G = (u_1, u_2, \ldots u_n)$. Recommendations are given to a user based on evaluation of items $I = (i_1, i_2, \ldots i_n)$ by other users from the same group (members of the group share common preferences). Therefore, one consider users from G who expressed their preferences regarding some of the items from I. Preferences can be expressed by grading or assigning some other value to the item, but it can also be extracted from the fact that users bought some item or have visited item details page many times. In order to make predictions these types of algorithms require user-item database and appropriate statistical techniques which can find the most similar or neighbor users. Finding the most similar users or neighbors is nothing more than calculating the correlation between system users.

Correlation coefficients are used to measure the strength of the relation between data (users, items, variables, etc.). There are several types of correlation coefficient and one of the most widely used is known as Pearson product moment correlation (PPMC) or in short Pearson's correlation:

$$sim\left(u_a, u_b\right) = \frac{\sum_{i_r \in IR}\left(r_{u_a,i} - \bar{r}_{u_a}\right)\left(r_{u_b,i} - \bar{r}_{u_b}\right)}{\sqrt{\sum_{i_r \in IR}\left(r_{u_a,i} - \bar{r}_{u_a}\right)^2}\sqrt{\sum_{i_r \in IR}\left(r_{u_b,i} - \bar{r}_{u_b}\right)^2}}, \qquad (8.8)$$

where:

– u_a, u_b—system users a and b, where $a, b \in U$,
– I, R—set of rated items, $i_r \in IR$.
– $r_{u_a,i}$—rating of user u_a for item i_r,
– $\bar{r}_{u_a}$—average rating of user u_a.

A correlation coefficient of 1 means that for every positive increase in one variable (user attribute), there is a positive increase of a fixed proportion in the other. For example, it is expected that with the increasing popularity of an author, the readership of his books should increase and vice versa. Zero value of correlation coefficient means that for every increase, there is not a positive or negative increase or in other words the two users are not very similar. High correlation value is preferred in order to combine preferences of similar users and create predictions or recommendations. Since the Pearson's correlation can result in positive and negative values, the absolute value (for example, $|-0.82| = 0.82$) gives us the relationship or similarity strength.

After determining the similarity between the users, by using Eq. (8.8), it is possible to calculate the predictions of the importance of an item for the observed user.

$$pred\,(u_a, i) = \bar{r}_{u_a} + \frac{\sum_{u_n \in U} sim\,(u_a, u_n)\,\left(r_{u_n,i} - \bar{r}_{u_n}\right)\left(r_{u_b,i} - \bar{r}_{u_b}\right)}{\sum_{u_n \in U} sim\,(u_a, u_n)} \qquad (8.9)$$

Item-based CF is also known as a model-based algorithm for making recommendations. Within this algorithm the similarities between different items in the set are calculated by using one of several similarity measures, and then these similarity values are used to predict ratings for user-item pairs not present in the dataset. As discussed in [21] *item-based* approach looks into the set of items the target user has rated and computes how similar they are to the target item i_t and then selects k most similar items $\{i_1, i_2, \ldots i_n\}$. Similarity between items can be calculated by using cosine similarity shown in Eq. (8.10), which measures the similarity between two vectors of an inner product space. It is measured by the cosine of the angle between two vectors and determines whether two vectors are pointing in roughly the same direction. It is often used to measure document similarity in text analysis where a document can be represented by thousands of attributes, each recording the frequency of a particular word (such as a keyword) or phrase in the document. This can also be applied to similarity between books or any other item, it is only important that its attribute values can be adequately described using a vector:

$$sim\,(\mathbf{i_1}, \mathbf{i_2}) = \frac{\mathbf{i_1} \cdot \mathbf{i_2}}{|\mathbf{i_1}|\,|\mathbf{i_2}|}, \qquad (8.10)$$

where:

- $\mathbf{i_1} \cdot \mathbf{i_2} = \sum_{f=1}^{n} i_{1f} i_{2f} = i_{11} i_{21} + i_{12} i_{22} + \ldots + i_{1n} i_{2n}$,
- $|\mathbf{i_1}| = \sqrt{i_{11}^2 + i_{12}^2 + \ldots + i_{1n}^2}$.

It is important to mention that CF is not immune to problems such as the cold start. If there is no user rated items or a new item is added, the system will not be able to generate appropriate recommendations using CF. Until a new item is rated by a sufficient number of users, the system will not recommend it. This problem is especially pronounced with the system where there exists constant input of new items. Since CF is based on system knowledge of what users prefer, the recommendations depend entirely on what content the users are evaluating. Users whose interests differ from other participants and for which it is difficult to find similar users cannot expect good results from CF. Therefore, CF has proven successful for those users who can be classified as the closest neighbors (clusters), while it is not possible to offer appropriate recommendations to users who do not have close neighbors. As discussed in [22] this leads to a sparse user-item matrix, inability to locate successful neighbors resulting in the generation of weak recommendations.

Model-Based Filtering Technique

Model-based approach intends to solve the problems issued by memory-based approach regarding the responsiveness to user requests. Model-based filtering technique uses usually clustering and classification, e.g., using Bayesian or neural networks for CF. This approach tries to compress huge database into a single model and to perform recommendation tasks by applying reference mechanism into this model (see [20]).

Clustering CF, described by Ungar et al. [23] resides on assumption that users in the same group have the same interest; so they rate items similarly. Therefore users are partitioned into groups called clusters which are defined as a set of similar users. Suppose each user is represented as rating vector denoted $u_i = (r_{i1}, r_{i2}, \ldots r_{in})$. The dissimilarity measure between two users is the distance between them. We can use Minkowski distance equation (8.11), Euclidian distance equation (8.5), or Manhattan distance equation (8.12).

$$D\left(X, Y\right) = \sqrt[p]{\left(\sum_i^n |x_i - y_i|^p\right)} \tag{8.11}$$

$$d_M\left(X, Y\right) = \sum_i^n |x_i - y_i|. \tag{8.12}$$

Classification CF foresees that each user is represented as rating vector $u_i = (r_{i1}, r_{i2}, \ldots r_{in})$.

Let us suppose that every rating value r_{ij}, which is an integer, ranges from c_1 to c_m. For instance, in a 5-values rating system, there are $c_1 = 1, c_2 = 2, c_3 = 3, c_4 = 4$, and $c_5 = 5$. Rating 5 given to an item means that such item is most favored while rating 1 to an item means such item is least favored by the user. Each c_k is considered as a class and the set $C = c_1, c_2, \ldots, c_m$ is known as class set.

From Table 8.2, active user vector $u_4 = (1, 2, ?, ?)$ has two missing value r_{43} and r_{44}. According to classification CF, predicting values of r_{43} and r_{44} is to find classes of r_{43} and r_{44} with suppose that there are only five classes $c_1 = 1, c_2 = 2, c_3 = 3, c_4 = 4, c_5 = 5$ in Table 8.2.

A popular classification technique is naïve Bayesian method, in which the user u_i belongs to class C if the posterior conditional probability of class C given user u_i is maximal [24]:

$$c = argmax(P(c_k|u_i)), c_k \subset C. \tag{8.13}$$

Table 8.2 Rating matrix
(user 4 is active user) as in
[20]

	Item 1	Item 2	Item 3	Item 4
User 1	$r_{11} = 1$	$r_{12} = 2$	$r_{13} = 1$	$r_{14} = 5$
User 2	$r_{21} = 2$	$r_{22} = 1$	$r_{23} = 2$	$r_{24} = 4$
User 3	$r_{31} = 4$	$r_{32} = 1$	$r_{33} = 5$	$r_{34} = 5$
User 4	$r_{41} = 1$	$r_{42} = 2$	$r_{43} = ?$	$r_{44} = ?$

8.1.5   Hybrid Recommender Systems

Most of the algorithms used to generate some form of recommendation have certain limitations and disadvantages. Ever since the advent of the first recommendation algorithms, efforts have been made to combine them in order to eliminate individual limitations by acting together. Mixed or hybrid recommendation systems combine the generation of content-based recommendations and collaborative filtering in order to overcome limitations of individual approaches, considering the character- istics of the content and the evaluation of the content by the user. Implementation approaches can be classified based on the different methods in order to generate recommendation. Some of the options are to modify the methods used in several stages based on a specific criterion or using one method to produce a model that will be used as input for other methods. Also, a slightly simpler form would imply to jointly present recommendations derived from different techniques or methods. Research describes [25] different combination of methods that have been employed, and some of them are: weighted, switching, mixed, feature combination, cascade, feature augmentation, and meta-level. A weighted hybrid recommender is one in which the score of a recommended item is computed from the results of all recommendation techniques present in the system. As discussed in [26] weighted technique computes the prediction score by considering them as variables in a linear combination. This technique gives each of them weight and summing up the weighted results.

Supposed that there are n recommendation approaches to be combined using weighted strategy, the prediction score ps of user u to item i can be computed by using:

$$ps_{u,i} = \sum_{j}^{n} \sigma_j ps_{u,i}^{(j)}, \qquad (8.14)$$

where,

- σ_j denotes weight of algorithm $ps_{u,i}$.

If only two recommendation approaches have been combined, then $n = 2$ and prediction score can be computed by using (8.15):

$$ps_{u,i} = \sigma_1 \cdot ps_{u,i}^{(1)} + (1 - \sigma_1) \cdot ps_{u,i}^{(2)}. \qquad (8.15)$$

Switching hybrid uses some criteria to switch between recommendation techniques in a way that the system uses a content/collaborative hybrid in which a content-based recommendation method is employed first. If the content-based system cannot make a recommendation with enough confidence, then a collaborative recommendation is attempted. Other researches [27] have proposed a switching hybrid recommendation approach by combining item-based collaborative filtering with a classification approach. They empirically showed that their recommendation approach outplays others in terms of accuracy, and coverage and scalability.

Mixed hybrid approach can be used in cases where it is practical to make large number of recommendations simultaneously. This implies that recommendations from more than one technique are presented together. It is important to emphasize that the mixed hybrid avoids the new item start-up problem in a way that the content-based component can recommend new items based on their descriptions even if they have not been rated by anyone. However mixed hybrid approach does not get around the new user start-up problem, since both the content and collaborative methods need some data about user preferences in order to produce appropriate recommendations [26].

In *Feature combination* the features from different recommendation data sources are used together into a single recommendation algorithm. Feature combination hybrid lets the system consider collaborative data without relying on it exclusively, so it reduces the sensitivity of the system to the number of users who have rated an item. Conversely, it lets the system have information about the inherent similarity of items that are otherwise opaque to a collaborative system [28]. In [29] authors proposed a feature combination hybrid that combined collaborative features such as user's likes and dislikes with content features of catalog items. As a result, they identified new features like users who like dramas to determine similar peers within the community. Feature augmentation is a strategy for hybrid recommendation that is similar in some ways to feature combination. Instead of using features drawn from the contributing recommender's domain, a feature augmentation hybrid generates a new feature for each item by using the recommendation logic of the contributing domain [30].

Cascade hybrid creates a strictly hierarchical hybrid where weak recommender cannot overturn decisions made by a stronger one but can merely refine them. It is similar to the feature augmentation hybrid, but it is an approach that retains the function of the recommendation component as providing predicted ratings. A cascade recommender uses a secondary recommender only to break ties in the scoring of the primary one [30].

Meta-level combination indicates the possibility where recommendation techniques are combined in a way that the model generated by one technique represents the input for another. The benefit of the meta-level method especially for the content/collaborative hybrid is that the learned model is a compressed representation of a user's interest, and a collaborative mechanism that follows can operate on this information-dense representation more easily than on raw rating data. Research [31] proposed collaborative constraint-based recommender system as a meta-level hybrid that combines collaborative filtering with the knowledge-based recommendation.

The collaborative filtering step takes the original community data as well as the current user model and the generated constraint candidates as input and selects a set of constraints that serves as a knowledge base for the subsequent recommendation step.

8.2   Building Recommender System Using ML.NET

ML.NET is an open source framework developed by Microsoft for training, building, evaluation, and deployment of machine learning models on .NET based applications [32]. It allows users to deliver custom machine learning models using C# or F# programming languages for various ML problems, e.g., regression, classification, sentiment analysis, time series forecasting, recommendations, image recognitions, on various software platforms, e.g., desktop, web, mobile, and cloud solutions. The framework is built on the pipeline pattern which is ideal for developing machine learning models. In the pipeline pattern, processing elements are arranged so that the output of each element is the input of the next, which was found in analogy to a physical pipeline [33]. Pipeline pattern for developing machine learning libraries are proved to be very suitable, and it has been already used in similar .NET projects [34].

Usually common ML projects consist of predefined ordered steps (Fig. 8.5):

- define a problem and collect the data related to the problem,
- load the data into a `DataView` object,
- define set of pipeline pattern based operations related to data transformation,
- define set of features and label, for a ML algorithm,
- train a model by calling `Fit()` on the pipeline,
- perform set of statistical estimations for model evaluation,
- in case the model is not satisfied the performance expectation, steps 3,4, 5, and 6 are repeated,
- persist the model as file, for use in an application,
- load the model into an `ITransformer` object for prediction,
- make predictions by calling `CreatePredictionEngine.Predict()`.

8.2.1   Data Preparation in ML.NET

Once the problem is defined and the decision of using machine learning technique has been approved, the first step in ML model development is data preparation. Data preparation is the key for successful ML application and plays the central role of the ML model life cycle. Often, the data that described the problem is not suitable to be used directly for ML algorithm, because they must be prepared in a table form

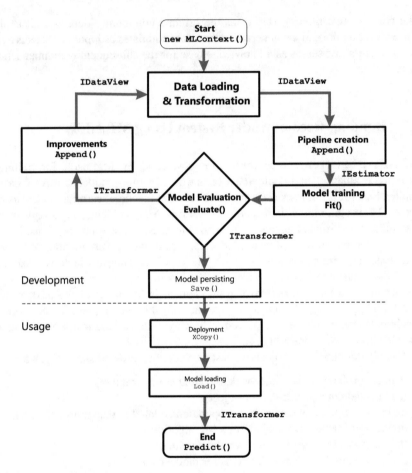

Fig. 8.5 ML model lifecycle workflow

where all columns are of numeric types. There are several problems when preparing data for machine learning algorithm:

- transformation text and categorical data into their numeric representation,
- handling missing values, and replacing them with meaningful values,
- feature engineering and selection,
- data normalization and scaling,
- custom transformation.

The ML.NET has built-in pipeline-based functionality for most of the data preparation, but sometimes the developer is required to include his custom data preparation modules. Data transformations can be chained together into pipeline so that each transformation both expects and produces data of specific types and formats. In ML.NET all data transformation implement the IEstimator interface [32].

8.2.2 Transformation Text Based Data

The transformations text data is one of the most common tasks in ML application. When processing text data usually one of following tasks are used: *tokenization* for converting sentences to words, and/or removing unnecessary punctuation and tags, *removing stop words*, and *stemming and lemmatization* in order to reduce different variation of the same word like studies, studying to a common base form study. Once the text data has been transformed into words, they became categorical data, which need to be encoded into numerical representation. One of the common techniques to numerically represent text is *Bag of Words*.

Some data transformations require training data to calculate their parameters. For example, the *NormalizeMeanVariance* transformer calculates the mean and variance of the training data during the Fit() operation, and uses those parameters in the *Transform()* operation.

The ML.NET has built-in functionality to transform text-based data into ML ready feature values. Beside text, ML.NET also contains the implementations for encoding categorical data into various types of numerical representations, e.g., label encoding, one-hot encoding, binary encoding, etc.

Handling missing values in dataset is important because your implementation will be crashed or at least will break the training process. So it is very important to check the training data against the missing values before the data get into the ML algorithm. Depending on the type of the missing values, handling them can be achieved in different ways. In case there are only few values that are missing in the dataset, the rows with missing values can be just removed. For numerical types, the missing values can be replaced with one of the parameters of descriptive statistics of the corresponded column. In other words, missing values can be replacing with maximum, or minimum, or average column value. Replacing the categorical values follows different approaches [35].

In ML.NET, central role of the data preparation is defined by the `IDataView` interface with implementation of the `DataView` class. The `DataView` class implements various methods used for data manipulation and preparation with abstractions similar to relational database. It provides compositional processing of schematized data with full support of the pipeline pattern, with generics abstraction of primitive operators and the composition of multiple operators in order to achieve higher-level semantics such as the `FeaturizeText` transform. The operators implemented, based on the `IDataView` interface are able to gracefully and efficiently handle high-dimensional and large datasets with cursoring supports the well-known iterator model of databases [8].

8.3   Building Recommender System on Restaurant Dataset

In this chapter the recommender system was built for restaurant dataset using ML.NET. The dataset was created in the time period from Jul-2010 to Feb-2011. During that period users added and rated new and existing restaurants by filling the *21 attributes* from which 19 of them are provided by the user when he/she signs into the system. Data comprises *138 unique users* who contributed with information about 130 restaurants and accumulated 1161 ratings. Possible rating values are 0, 1, and 2, where 0 indicates that the user does not like the restaurant, and 2 denotes a high preference [9]. The recommender systems try to predict user's rating for the unrated restaurants. In order to build recommendation engine, the selected algorithm was field-aware factorization machines [9]. The algorithm uses *collaborative filtering* for recommendations which uses the fact that if a person A has the same opinion as a person B on the same restaurant, A is more likely to have B's opinion on a different restaurant than that of a random person.

8.3.1   Dataset Preparation

As previously mentioned, the dataset collects the users rating for the restaurants in the specified time period. The dataset was divided into two sets: the training set which holds the 80% of the original data, and the test set which holds the remaining 20% of the data. The user table contains 19 users attributes: *userID, latitude, longitude, smoker, drink level, dress preference, ambience, transport, marital status, hijos, birth year, interest, personality, religion, activity, color, weight, budget,* and *height.* The restaurant dataset describes each restaurant with *23 attributes.* Some of them are: *cuisine, alcohol, smoking, dress, accepts* (type of payment), *parking,* etc. The values were generated by users from several possible options for each attribute. The whole dataset can be viewed from the official site at http://kaggle. com/uciml. The last dataset used in the recommender system is the rating table which collects the users, restaurant and user's rating for the specific restaurant. The rating table contains the rating values (0,1 and 2) of all rated restaurants. The rating table contains 1161 ratings. Since the rating values can take 0, 1, or 2 value, the whole rating column transforms into a new binary column called recommended. The binary column indicates if the specific restaurant can be recommended for the specified user or not. Since the column values are mostly textual data, the data should be transformed into their numerical transformation by using several techniques which can be seen in the literature [7].

Table 8.3 FFM parameters used for model training

Parameter name	Parameter value
Learning rate	0.013
Number of iterations	450
Shuffle	True
Extra feature columns	*Smoking area, accessibility, dress code*
Extra feature columns	True

Table 8.4 Performance values for the restaurant recommender model

Parameter name	Parameter value
Accuracy	71.96%
Area under curve, AUC	82.10%
Area under precision/recall curve, ROC	74.82%
F1 score	69.39%
Log loss	0.82
Log loss reduction	0.18
Positive precision	0.68
Positive recall	0.71
Negative precision	0.75
Negative recall	72.88%

8.3.2   Model Training and Evaluation

Once the data is prepared the training phase consisted of the ML algorithm selection and calling the Fit() method by passing the transformed training dataset. The selected algorithm was *Field-aware Factorization Machines*, FFM which is proved to be very effective for building recommendation based on the multiple features [9]. Table 8.3 shows the parameters used during training of the recommender model.

Once the model is trained and persisted on the disk, the model evaluation was performed by calculation of several performance parameters for test dataset. The model evaluation shows that total accuracy for the test data is 73%, and the AUC = 0.83, and ROC = 74.5. The rest of the parameters were shown in Table 8.4. In case of random binary classifier, the ROC and AUC values are 50%, which is much lower than performance of the recommender model presented in the paper.

8.3.3   Integration of ML Models in the Cloud Based Applications

Integration ML models into cloud based applications can be achieved by several ways. One of the ways is by using APIs under Microsoft.ML.Extension namespace of ML.NET Framework. This part of the APIs provides standardized way to integrate loading ML.NET model for scoring in various cloud based apps, e.g., ASP.NET apps, Azure Functions, or web services. Using dependency injection, the extension APIs of the ML.NET provides a developers a well know software technology for integration and optimizing the model's execution and performance in multi-threaded environments such as ASP.NET Core applications. Besides using APIs directly from the framework, ML models can be integrated into the cloud environment by using different techniques proposed at the literature [36, 37].

8.4 Conclusion

The chapter presents systematic overview of the main types of recommender systems. The overview shown that recommender systems can be implemented on various ways depending of the user purpose, and application. On the other hand, the application of the recommender systems requires to use proven and quality software library which is mostly independent of the platform so the ML solution can be easily implemented and delivered. One of recently developed framework and quickly became popular is Microsoft open source framework for machine learning called ML.NET. The chapter also describes the methodology and use case scenario how to use ML.NET an open source framework in building recommend systems in general, with application for the restaurants. Through the several phases the recommender system was built by providing details on each phase. The chapter presented how to load, prepare, and transform the data. Data transformation also included creation of the training and testing datasets, so that after training phase the model evaluation can provide reliable decision whether the model can be used in the production. Through the example, the model of the recommender system was successfully built using ML.NET platform and FFM algorithm. Once developed it can be integrated as a part of the cloud solution in many ways using different approaches.

References

1. D. Goldberg, D. Nichols, B.M. Oki, D. Terry, Using collaborative filtering to weave an information tapestry. Commun. ACM **35**(12), 61–70 (1992). https://doi.org/10.1145/138859. 138867
2. P. Resnick, N. Iacovou, M. Suchak, P. Bergstrom, J. Riedl, GroupLens: an open architecture for collaborative filtering of netnews, in *Proceedings of the 1994 ACM Conference on Computer Supported Cooperative Work (CSCW '94)* (ACM, New York, 1994), pp. 175–186. https://doi.org/10.1145/192844.192905
3. U. Shardanand, P. Maes, Social information filtering: algorithms for automating "Word of Mouth", in *CHI* (1995)
4. W. Hill, L. Stead, M. Rosenstein, G. Furnas, Recommending and evaluating choices in a virtual community of use, in *Proceedings of the SIGCHI Conference on Human Factors in Computing Systems (CHI '95)*, ed. by I.R. Katz, R. Mack, L. Marks, M.B. Rosson, J. Nielsen (ACM Press/Addison-Wesley Publishing, New York, 1995), pp. 194–201. https://doi.org/10.1145/223904.223929

5. G.D. Linden, J.A. Jacobi, E.A. Benson, Collaborative recommendations using item-to-item similarity mappings. US Patent 6266649 (2001). https://doi.org/10.1145/584792.584803. ISBN=1581134924
6. J. Davidson, B. Liebald, J. Liu, P. Nandy, T. Van Vleet, U. Gargi, S. Gupta, Y. He, M. Lambert, B. Livingston, D. Sampath, The YouTube video recommendation system, in *Proceedings of the Fourth ACM Conference on Recommender Systems (RecSys '10)* (ACM, New York, 2010), pp. 293–296. https://doi.org/10.1145/1864708.1864770
7. C.A. Gomez-Uribe, N. Hunt, The Netflix recommender system: algorithms, business value, and innovation. ACM Trans. Manag. Inf. Syst. **6**(4), 19 pages, Article 13 (2015). https://doi.org/10.1145/2843948
8. B. Smith, G. Linden, Two decades of recommender systems at Amazon.com. IEEE Internet Comput (2017). ISSN = 10897801. https://doi.org/10.1109/MIC.2017.72
9. https://www.netflixprize.com/index.html. Accessed 22 June 2019
10. E.-A. Baatarjav, S. Phithakkitnukoon, R. Dantu, Group recommendation system for Facebook, in *Proceedings of the OTM Confederated International Workshops and Posters on On the Move to Meaningful Internet Systems: 2008 Workshops: ADI, AWeSoMe, COMBEK, EI2N, IWSSA, MONET, OnToContent + QSI, ORM, PerSys, RDDS, SEMELS, and SWWS (OTM '08)*, ed. by R. Meersman, Z. Tari, P. Herrero (Springer, Berlin, 2008), pp. 211–219. https://doi.org/10.1007/978-3-540-88875-8-41
11. O. Phelan, K. McCarthy, B. Smyth, Using Twitter to recommend real-time topical news, in *Proceedings of the Third ACM Conference on Recommender Systems (RecSys '09)* (ACM, New York, 2009), pp. 385–388. https://doi.org/10.1145/1639714.1639794
12. B.T. Betru, C.A. Onana, B. Batchakui, Deep learning methods on recommender system: a survey of state-of-the-art. Int. J. Comput. Appl. (2017). https://doi.org/10.5120/ijca2017913361
13. S. Zhang, L. Yao, A. Sun, Y. Tay, Deep learning based recommender system: a survey and new perspectives. ACM Comput. Surv. **52**(1), 38 pages, Article 5 (2019). https://doi.org/10.1145/3285029
14. G. Preethi, P.V. Krishna, M.S. Obaidat, V. Saritha, S. Yenduri, Application of deep learning to sentiment analysis for recommender system on cloud, in *IEEE CITS 2017–2017 International Conference on Computer, Information and Telecommunication Systems* (2017). https://doi.org/10.1109/CITS.2017.8035341. ISBN = 9781509059560
15. R. Salakhutdinov, A. Mnih, G. Hinton, Restricted Boltzmann machines for collaborative filtering, in *Proceedings of the 24th International Conference on Machine Learning (ICML '07)*, ed. by Zoubin Ghahramani (ACM, New York, 2007), pp. 791–798. https://doi.org/10.1145/1273496.1273596
16. A. van den Oord, S. Dieleman, B. Schrauwen, Deep content-based music recommendation, in *Proceedings of the 26th International Conference on Neural Information Processing Systems (NIPS'13)*, vol. 2, ed. by C.J.C. Burges, L. Bottou, M. Welling, Z. Ghahramani, K.Q. Weinberger (Curran Associates, New York, 2013), pp. 2643–2651
17. D. Jannach, M. Zanker, A. Felfernig, G. Friedrich, *Recommender Systems: An Introduction*, 1st edn. (Cambridge University Press, New York, 2010)
18. C.C. Aggarwal, *Recommender Systems: The Textbook*, 1st edn. (Springer Publishing Company, Incorporated., New York, 2016). ISBN = 9781605580937. https://doi.org/10.1145/245108.245121
19. A.I. Schein, A. Popescul, L.H. Ungar, D.M. Pennock, Methods and metrics for cold-start recommendations, in *Proceedings of the 25th Annual International ACM SIGIR Conference on Research and Development in Information Retrieval (SIGIR '02)* (ACM, New York, 2002), pp. 253–260. https://doi.org/10.1145/564376.564421
20. M.-P.T. Do, D. Van Nguyen, L. Nguyen, Model-based approach for collaborative filtering, in *6th International Conference on Information Technology for Education* (2010)
21. B. Sarwar, G. Karypis, J. Konstan, J. Riedl, Item-based collaborative filtering recommendation algorithms, in *Proceedings of the 10th International Conference on World Wide Web (WWW '01)* (ACM, New York, 2001), pp. 285–295. https://doi.org/10.1145/371920.372071

22. P. Aggarwal, V. Tomar, A. Kathuria, Comparing content based and collaborative filtering in recommender systems. Int. J. New Technol. Res. (IJNTR) **3**(4) (2017). ISSN = 2454-4116

23. L. Ungar, D. Foster, E. Andre, *Star Wars and Fred Star Wars and Dean Star Wars and Jason Hiver Whispers. Clustering Methods for Collaborative Filtering* (AAAI Press, Palo Alto, 1998)

24. J. Han, M. Kamber, J. Pei, *Data Mining Concepts and Techniques*, 3rd edn. (Morgan Kaufmann, Waltham, 2012). ISBN = 0123814790

25. R. Burke, Hybrid recommender systems: survey and experiments. User Model. User-Adap. Inter. **12** (2002). ISSN = 1573-1391. https://doi.org/10.1023/A:1021240730564

26. S. Suriati, M. Dwiastuti, T. Tulus, Weighted hybrid technique for recommender system. J. Phys. Conf. Ser. **930** (2017). https://doi.org/10.1088/1742-6596/930/1/012050

27. M.A. Ghazanfar, A. Prugel-Bennett, Building switching hybrid recommender system using machine learning classifiers and collaborative filtering. IAENG Int. J. Comput. Sci. **37**(3), 1–16 (2010). ISSN = 1819656X

28. J. Shah, L. Sahu, A survey of various hybrid based recommendation method. Int. J. Adv. Res. Comput. Sci. Softw. Eng. **4**(11), 367–371 (2014). ISSN=2277 128X

29. C. Basu, H. Hirsh, W. Cohen, Recommendation as classification: using social and content-based information in recommendation, in *Proceedings of the Fifteenth National/Tenth Conference on Artificial Intelligence/Innovative Applications of Artificial Intelligence (AAAI '98/IAAI '98)* (American Association for Artificial Intelligence, Menlo Park, 1998), pp. 714–720

30. E. Cano, M. Morisio, Hybrid recommender systems: a systematic literature review. Intell. Data Anal. (2017). ISSN = 15714128. https://doi.org/10.3233/IDA-163209

31. M. Zanker, A collaborative constraint-based meta-level recommender, in *Proceedings of the 2008 ACM Conference on Recommender Systems* (ACM, New York, 2008). ISBN = 978-1-60558-093-7

32. Z. Ahmed, S. Amizadeh, M. Bilenko, R. Carr, W.-S. Chin, Y. Dekel, X. Dupre, V. Eksarevskiy, S. Filipi, T. Finley, A. Goswami, M. Hoover, S. Inglis, M. Interlandi, N. Kazmi, G. Krivosheev, P. Luferenko, I. Matantsev, S. Matusevych, S. Moradi, G. Nazirov, J. Ormont, G. Oshri, A. Pagnoni, J. Parmar, P. Roy, M.Z. Siddiqui, M. Weimer, S. Zahirazami, Y. Zhu. Machine learning at Microsoft with ML.NET, in *Proceedings of the 25th ACM SIGKDD International Conference on Knowledge Discovery and Data Mining (KDD '19)* (ACM, New York, 2019), pp. 2448–2458. https://doi.org/10.1145/3292500.3330667

33. S. Vinoski, Chain of responsibility. IEEE Internet Comput. (2002). ISSN = 1089-7801. https://doi.org/10.1109/MIC.2002.1067742

34. https://github.com/UniversityOfAppliedSciencesFrankfurt/LearningApi. Accessed 22 June 2019

35. B. Hrnjica, A. Danandeh Mehr, Optimized genetic programming applications. IGI Global (2018). ISBN = 9781522560050. https://doi.org/10.4018/978-1-5225-6005-0

36. J. Zhao, T. Tiplea, R. Mortier, J. Crowcroft, L. Wang, Data analytics service composition and deployment on edge devices, in *Proceedings of the 2018 Workshop on Big Data Analytics and Machine Learning for Data Communication Networks (Big-DAMA '18)* (ACM, New York, 2018), pp. 27–32. https://doi.org/10.1145/3229607.3229615

37. B. Hrnjica, A. Danandeh Mehr, Energy demand forecasting using deep learning, in *Smart Cities Performability, Cognition, & Security* (Springer International Publishing, Cham, 2019). ISBN=978-3-030-14718-1. https://doi.org/10.1007/978-3-030-14718-1-4

38. S. Alabady, F. Al Turjman, S. Din, A novel security model for cooperative virtual networks in the IoT era. Int. J. Parallel Prog. (2018). https://doi.org/10.1007/s10766-018-0580-z

39. F. Al Turjman, S. Alturjman, Confidential smart-sensing framework in the IoT era. J. Supercomput. **74**(10), 5187–5198 (2018)

40. F. Al Turjman, Price-based data delivery framework for dynamic and pervasive IoT. Pervas. Mobile Comput. J. **42**, 299–316 (2017)

41. F. Al Turjman, M.Z. Hasan, H. Al Rizzo, Task scheduling in cloud-based survivability applications using swarm optimization in IoT. Trans. Emerg. Telecommun. **30**(8) (2019). https://doi.org/10.1002/ett.3539

42. F. Ricci, L. Rokach, B. Shapira, *Introduction to Recommender Systems Handbook* (Springer, Boston, 2011). ISBN = 978-0-387-85820-3. https://doi.org/10.1007/978-0-387-85820-3-1

Chapter 9
A Decade Bibliometric Analysis of Underwater Sensor Network Research on the Internet of Underwater Things: An African Perspective

Abdulazeez Femi Salami, Eustace M. Dogo, Tebogo Makaba, Emmanuel Adewale Adedokun, Muhammed Bashir Muazu, Bashir Olaniyi Sadiq, and Ahmed Tijani Salawudeen

9.1 Introduction

Recent advancements in cloud computing (CC), artificial intelligence (AI) and Internet of Things (IoT) have immensely revolutionized terrestrial wireless sensor networks (TWSN) communication which has consequently paved the way for the practical realization of underwater wireless sensor networks (UWSNs) with the objective of supporting the emergence of Internet of Underwater Things (IoUT) [1–6]. IoUT strategically applies IoT to water bodies through a specialized and dedicated network of smart coordinated underwater sensors for monitoring unexplored water regions in any geographical area of interest [7–12]. These digital underwater sensors have the ability to interpret and interact with the geographical area of interest for preserving aquatic habitats and managing underwater resources by taking advantage of equipment waterproofing techniques, embedded systems,

A. F. Salami (✉)
Department of Computer Engineering, University of Ilorin, Ilorin, Nigeria
e-mail: salami.af@unilorin.edu.ng

E. M. Dogo
Department of Electrical and Electronic Engineering Science, Institute for Intelligent Systems, University of Johannesburg, Johannesburg, South Africa
e-mail: eustaced@uj.ac.za

T. Makaba
Department of Applied Information Systems, University of Johannesburg, Johannesburg, South Africa
e-mail: tmakaba@uj.ac.za

E. A. Adedokun · M. B. Muazu · B. O. Sadiq · A. T. Salawudeen
Department of Computer Engineering, Ahmadu Bello University, Zaria, Nigeria
e-mail: wale@abu.edu.ng; mbmuazu@abu.edu.ng; bosadiq@abu.edu.ng; tasalawudeen@abu.edu.ng

© Springer Nature Switzerland AG 2020
F. Al-Turjman (ed.), *Trends in Cloud-based IoT*, EAI/Springer Innovations in Communication and Computing, https://doi.org/10.1007/978-3-030-40037-8_9

intelligent controllers and actuators, powerful tracking devices and Internet technology to generate real-time data that is available and accessible through different communication modes such as Thing to Thing (T2T) or Human to Thing (H2T) [13–15].

Due to the aforementioned capacity of underwater sensor technologies, IoUT are considered as an integral and critical component for building smart cities (SC) [6, 18–21]. UWSN is also an indispensable and vital technological asset for SC as about 70% of the earth is estimated to be covered by water bodies, and it is evident that oceans to a large extent regulate global climatic conditions and wind patterns that seriously affect terrestrial life [1, 8, 18–20, 22]. This inarguably implies that Internet of Terrestrial WSN can potentially only connect around 30% of the earth [23–26]. UWSN employs a fusion of vehicular, wireless and micro-electromechanical sensor technology equipped with intelligent computing/processing, interactive communication/interconnection and smart sensing/detection capabilities in order to achieve coordinated event (water-related properties and conditions) observation and data aggregation tasks through submerged sensor nodes, surface gateway nodes and coastal control station [24, 27–33]. These sensed underwater data are further utilized and processed by specific end-user applications in order to cater for environmental and human needs [34–40].

The need for better environmental monitoring within the context of smart cities and the recent spate of global natural disasters has further aroused research interest in IoUT [41–46]. This is motivated by a number of UWSN innovations, such as the development of tethered remotely operated underwater vehicles (ROUVs), untethered autonomous underwater vehicles (AUVs), unmanned/autonomous surface vehicles (USVs/ASVs) and other smart underwater technologies [33, 35, 37, 38, 47–54]. While these inventions hold promising prospects for technologically advanced countries, the same assertion cannot be made for most African countries due to challenges inherent in research and development (R&D) activities into critical IoUT/UWSN projects in the region [18, 19, 44, 55].

The objective of this chapter is to conduct a bibliometric analysis that systematically highlights the knowledge base for core research works in UWSN within the African region. The chapter employs VOSviewer science mapping software tool to analyse scientific research journals published in SCOPUS-indexed database over a period of 10 years (2008–2019) in the field of UWSN.

This research methodically investigates and interprets the ensuing findings of the analysis in order to carve out useful technical approaches and important procedures for exploiting the knowledge domain and intellectual structure of UWSN research within the African context. Furthermore, this analysis identifies and highlights vital missing links, essential research directions and unique technical contributions and recommendations that will be of relevance in facilitating the successful actualization of IoUT/UWSN research projects in Africa.

Section 9.2 of this chapter provides a technical overview of UWSN together with its conceptual definition, salient features and architecture. The potentials, applications and use case scenarios for UWSN are also discussed in this section. Additionally, the paradigm of UWSN in the cloud is also covered in this section together with pertinent design challenges and practical solutions. In Sect. 9.3,

IoUT is introduced together with its technical definitions, key characteristics and architecture. Potential applications and implementation instances for IoUT are also outlined in this section. Furthermore, the concept of cloud-based IoUT is outlined together with the associated implementation issues and relevant solutions. Data extraction, analysis and other statistical techniques employed with respect to the methodology of the research conducted in this chapter are discussed in Sect. 9.4. Section 9.5 presents the ensuing results upon conducting comprehensive analysis, Sect. 9.6 explicates the findings of this analysis while the conclusion of this chapter is given in Sect. 9.7.

9.2 Underwater Sensor Networks and the Internet of Underwater Things

9.2.1 Technical Definitions, Features, Architecture and Communication Capabilities

IoUT is the strategic application IoT to water bodies through a specialized network of smart interlinked underwater sensors for monitoring vast water regions in a geographical area of interest [7–12, 37, 38]. These digital underwater sensors have the ability to interpret, interact and adapt with the underwater environment in order to preserve aquatic habitats and manage underwater resources by taking advantage of equipment waterproofing techniques, embedded systems, intelligent controllers and actuators, powerful tracking devices and Internet technology [13–17, 56]. IoUT relies on the fusion of these technologies to deliver real-time and reliable data that is available and accessible through different communication modes such as T2T or H2T [15, 17, 56–58]. Due to the potentials of underwater sensor technologies, IoUT are considered as an essential and vital technology for building SC [6, 18, 19, 21, 59, 60].

The attributes of IoUT are, namely, sparse network density (difficult to establish communication and maintain connection among various IoUT devices), sound-emitting tracking technologies (acoustic tags have a relatively broader detection/sensing area of approximately 1 km in freshwater and can track aquatic targets in three-dimension), energy harvesting capabilities (ocean thermal energy, solar energy, piezoelectric schemes, microbial fuel cell techniques), and different localization and communication (usually acoustic) mechanisms [8, 10, 20, 61, 62].

A generic three-layered IoUT architecture for monitoring ocean acidification is depicted in Fig. 9.1. From the IoUT architecture, the perception or sensing layer has direct physical contact with the underwater environment [62]. In the sensing layer, aquatic data are gathered with the aid of various sensory objects (acoustic tags, AUVs, underwater sensors, surface sinks) [8, 62]. The network layer transmits the processed data from the sensing layer to a remote coastal control station through the fusion of CC, Internet, network and database management technologies [8, 10,

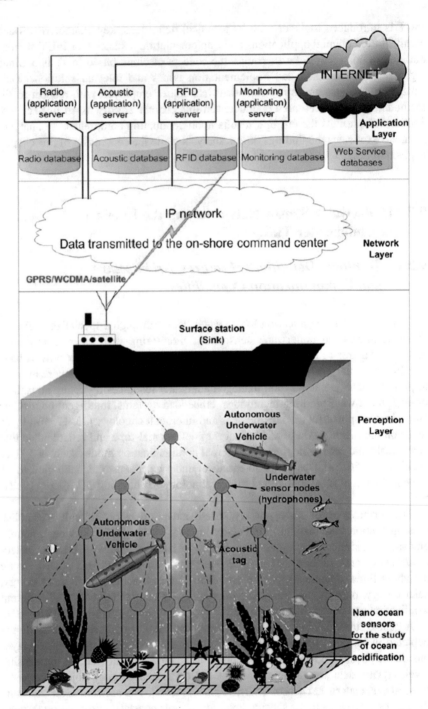

Fig. 9.1 IoUT Architecture [62]

62]. The application layer employs innovative AI techniques for big data analytics, behavioural analytics, data mining and advanced predictive analytics in order to generate vital information that will meet the demands of end users and dedicated systems [10, 62, 63].

UWSN is a specialized network of smart collaborative sensors deployed for monitoring aquatic habitats, exploring underwater resources, inspecting submerged assets, infrastructures and installations, and collecting sub-aquatic data with respect to an event of interest [8, 20, 28, 30, 31]. It must be mentioned that these autonomous underwater sensors can be static (where nodes are anchored or attached to docks or seafloor), semi-mobile (where nodes are temporarily suspended from ship-deployed buoys) or mobile (where nodes are attached to gliders, drifters or floaters) and they are usually distributed in a deployment environment of deep water (where sensors attached to AUVs are employed) [8, 14, 20, 27, 33]. The aggregated sensed data is relayed to surface sinks (ASVs, gateway nodes mounted on a ship or strategically placed buoyant nodes), and afterwards the sinks forward desired data to a remote coastal control station where the data is intelligently analysed in order to extract useful, critical and vital information [14, 34, 36–38]. The corresponding UWSN architecture depicting this scenario is as shown in Fig. 9.2.

One of the peculiarities of UWSN is that underwater instruments are expensive and non-disposable [1, 8]. In addition to this, UWSN deployments are highly resource-intensive, harsher (requiring periodic maintenance), scheduled for shorter time periods (typically in hours) and targeted for longer range but with relatively less dense network (considerably fewer nodes) than TWSN [8, 14]. Furthermore, UWSN deployment and packaging costs are comparatively very high [8, 20]. Due to the fact that UWSN deployments are targeted for longer range, acoustic communication techniques are mostly employed as electrostatic, radio and optical signals attenuate very quickly within a short range of 1–10 m [8, 33, 64].

However, the performance of acoustic communications in underwater environments varies with water depth, with shallow water (exhibits high temperature) having a depth of 0–100 m while deep water (possesses low temperature) ranging from 100 m to 10,000 m [8, 23, 24, 33, 64]. Deep water is more favourable for acoustic communications while shallow water severely affects acoustic signals as a result of Doppler spread, surface noise (noise by human-beings and ambient noise), salinity, high temperature gradients, shadow zones, low sound speed, time-varying multipath and fading effects, path loss (attenuation, scattering loss and geometric spreading loss), variable and long propagation delays, high latency, narrow and distance-dependent bandwidth, node failures as a result of corrosion and bio-fouling, dynamic topology and frequent node mobility due to water current activities resulting in more energy consumption [8, 14, 20, 64].

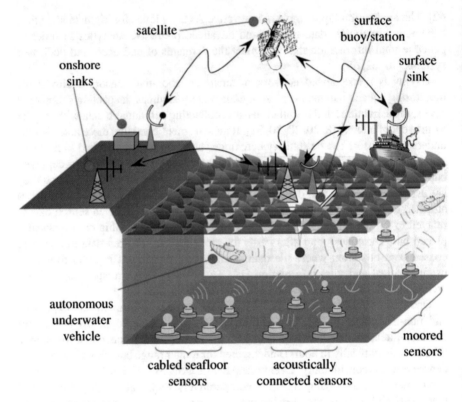

Fig. 9.2 UWSN architecture [8]

9.2.2 Applications and Use Case Scenarios in Africa

In Nigeria, Shell Petroleum Development Company (SPDC) in partnership with Fugro Group launched and operated two REMUS 100 AUVs (equipped with inertial navigation system and reconfigurable sensors with operational period of 6–8 h at a water depth of 100 m) for inspecting and surveying subsea assets (pipelines, gas gathering systems, mooring platforms, floating production storage and offloading vessels) off the Niger Delta Coastline (NDC) [65, 66]. These AUVs are also employed for detecting drowned wrecks, obstacles and rocks that can disrupt or hinder the safe and smooth setup of offshore facilities [66]. The monitoring and control centre was remotely located on the main operations vessel kept at a secure operating distance from the NDC [66]. These underwater sensors were employed as a result of the challenging security situation in the Niger Delta area and, most importantly, to reduce the risk of exposing offshore personnel to hazards and various forms of threatening situations [65, 66]. Moreover, the traditional approach of deploying a slow-moving vessel for surveying subsea assets is ineffective and

Fig. 9.3 AUV for subsea asset monitoring [65]

risky for a security-challenged region [66]. UWSN was therefore considered as an alternative surveying technique in order to preserve the integrity of subsea assets, maintain safe offshore operations and meet the growing demand for big data analytics on offshore assets' condition [65, 66]. Other practical benefits of adopting UWSN in this scenario are reduced time for assets monitoring (AUV takes around 20 s to turn while a traditional vessel takes approximately 30 min to make a turn), relatively shorter mobilization time (deployment time of AUV is considerably shorter than time needed to set up towed hydrophone arrays, knuckle-boom cranes, towed CTD chains, towed sub-bottom profilers, sonars and winch for a traditional vessel), easier and quicker access to high-quality and reliable dataset at a reduced cost [65, 66]. A pictorial view of this AUV is given in Fig. 9.3.

In Bamako, Mali, the River Basin Commission in cooperation with VIA Water, Indymo and Akvo Caddisfly deployed an underwater drone (fitted with sensors for bacteriological and physico-chemical pollution measurements) in the Niger River [67, 68]. The purpose of this deployment is to monitor and collect water quality data and publicly share this data on a website through the Internet with research agencies, universities, governmental bodies and local water consumers

Fig. 9.4 Underwater drone
for water quality monitoring
[67]

[67]. Another benefit of this IoUT/UWSN project is the ability to monitor previously inaccessible regions (interiors of drain and discharge pipes, remote shore boundaries and hard-to-reach portions of the River) [68]. This innovation allows the River Basin Commission to provide reliable real-time and continuous water quality data to all stakeholders (governmental bodies, policymakers, private organizations, fishermen, sand extractors, dyers) [67, 68]. This underwater drone is depicted in Fig. 9.4.

In Tanzania, Downstream Research and Conservation Limited in collaboration with Code for Africa installed and deployed underwater sensors to track illegal blast fishing activities [69]. Blast fishing with the use of dynamite or any homemade explosives for killing fishes destroys marine life and endangers the surrounding ecosystem by introducing pollutants, contaminants and other harmful substances [69]. The benefit of this IoUT/UWSN project is that important stakeholders (law enforcement agents, investigative journalists, local watchdogs) can leverage on the data from these underwater sensors to track and effectively curb illegal blast fishing activities [69]. This UWSN setup is as shown in Fig. 9.5.

9.2.3   Underwater Sensor Networks in the Cloud

In a cloud-based UWSN, knowledge is extracted from the received data in order to generate vital information that will enable remote end users (or systems) to properly understand, intelligently respond and adapt to the dynamic underwater environment

Fig. 9.5 Underwater acoustic sensors for tracking illegal blast fishing [69]

[18, 19, 55, 70]. In this scenario, core sensor-cloud management services are managed from the remote coastal control station [55, 70]. One of the benefits of cloud-based UWSN is the ability to cope with dynamic loads, minimize deployment cost, introduce novel supporting applications and, most importantly, improve the performance and accessibility of the UWSN [55, 70].

A three-layered architecture for cloud-based UWSN is shown in Fig. 9.6. This architecture consists of underwater sensor (US) which deals with aquatic data collection by underwater sensors, underwater web (UW) which employs sensor-cloud middleware (SCM) for managing large marine datasets and underwater data computing (UDC) which utilizes cloud interface services (CIS) and data cloud nodes (DCN) for aquatic data computation and analysis layers [70]. In this architecture, the embedded sensor-to-web gateway (ESWG) links the US and UW layers [70]. Global end users and systems (GEUS) interact with the cloud-based UWSN to have access to real-time aquatic data (for gradual predictive trends or unusual event demanding emergency response) through the UW graphical user interface (UW-GUI) [70].

9.2.4   Design/Implementation Challenges and Solutions

While the emerging field of UWSN is promising, there are some global challenges that need to be addressed in order to harness the full potentials of this technology [8, 10, 64]. These challenges are, namely, dynamic underwater environment (high pressure, uneven water surface depths, erratic underwater conditions), costly deployment and complex network design (frequent battery recharging, underwater

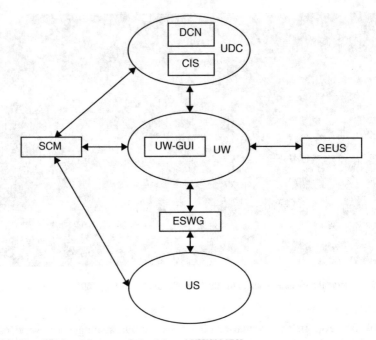

Fig. 9.6 Simplified architecture of cloud-based UWSN [70]

sensor recovery, hardware maintenance), scalability issues (vast underwater area needs devices, connections and installations that can measure up with the huge space), information reliability problems (localization and location where data is collected becomes problematic as water currents move nodes continuously and disrupts planned topology for data transmission), lack of backward compatibility with TWSN (communication medium, protocols and algorithms for TWSN cannot be extended directly to UWSN due to the large fundamental differences in their operational environment and communication mechanism), low bandwidth (water absorbs portion of transmission energy, takes time to transmit large data under such low frequencies), and hardware damages (mud and silt deposits from underwater erosion or flooding, corrosion from salt infiltration, spirogyra and other forms of algae accumulation on camera lens) [1, 8, 10, 14, 64].

9.3 Methodology

9.3.1 Data Extraction and Search Criteria

The search criteria used for this research span the period between 2008 through July 2019, using "Underwater sensor network" OR "Internet of Underwater Things" as search keywords query. This search was conducted on July 31, 2019. Initially, this research identified two sources of large academic databases: Elsevier's Scopus and Thomson Reuter's Web of Science (WoS). This research however narrowed its focus on documents published in Scopus due to its coverage and bias towards Engineering and Science-related disciplines [71]. Another motivating factor is that Scopus is reported to have a major strength and is a widely accepted repository in Engineering, Physical Science and Biomedical-related research disciplines [71] in comparison to its close competitor, the Web of Science (WoS). Additionally, Scopus has a larger database compared to WoS, which means more coverage than WoS as well as over 60% overlap of documents between the two databases. The choice of Scopus for this research is therefore informed by these empirical and comparative analyses of document coverage of the two indexed repositories. The preference for Scopus is also in line with the goal of this research which is to analyse broad interconnected research trends and gain empirical perspectives on UWSN/IoUT research over the past decade based on published scholarly articles, globally and in Africa. The search criteria employed in this research excluded articles published in other languages because majority of journals in Scopus are published in English language in comparison to other languages. Although this research has justified the use of Scopus-indexed publications for this review, this has led to the omission of non-Scopus-based publications. Hence, this research does not review the entire literatures in UWSN/IoUT domain. Nonetheless, the findings of this research, to a large extent, give a sufficient representation of research work in this field both globally and in Africa.

This research followed the reporting guidelines of PRIMA (Preferred Reporting Items for Systematic Reviews and Meta-Analyses) method [72], for identification and conducting systematic reviews in research. The following set of criteria was used in the Scopus search engine:

1. Inclusion date: 2008 to July 2019.
2. Inclusion document type: conference papers, articles, conference review, book, book chapter, review, editorial, letter and short survey as shown in Table 9.3.
3. Inclusion subject areas in Engineering and Science-related fields: Engineering, Energy, Chemical Engineering, Computer Science, Environmental Science, Decision Sciences, Earth and Planetary Sciences, Genetics and Molecular Biology, Biochemistry, Chemistry, Immunology and Microbiology. This information is given in Table 9.2.
4. Inclusion document source type: Conference Proceedings, Journals, Book Series, Books, Trade/Specialized Publications.
5. Excluded Languages: Chinese, French, and Portuguese.

Consequently, the Scopus search yielded a total of 1025 published articles between 2008 and July 31, 2019. The articles discovered from the document types and subject areas were further examined to ascertain their relevance. After the filtering, the dataset from the Scopus database was then exported into a Microsoft Excel.csv file. The data stored/collected contains the following information, namely, authors' name, authors' affiliation, articles' titles, abstracts, keywords and other key

Table 9.1 Top 20 Authors publishing in UWSN/IoUT

No.	Author	TNP	TNC	TNC/TNP
1	Cui, J.-H.	52	2187	28.20
2	Vieira, L.F.M	28	708	9.13
3	Peng, Z.	19	733	9.45
4	Coutinho, R.W.L.	18	352	4.54
5	Boukerche, A.	17	339	4.37
6	Javaid, N.	17	175	2.26
7	Petrioli, C.	17	297	3.83
8	Petroccia, R.	17	266	3.43
9	Loureiro, A.A.F.	16	369	4.76
10	Ammar, R.	15	106	1.37
11	Boutaleb, T.	14	96	1.24
12	Ghoreyshi, S.M.	14	96	1.24
13	Shahrabi, A.	14	96	1.24
14	Zhou, Z.	14	960	12.38
15	Guo, Y.	12	47	0.61
16	Liu, J.	12	307	3.96
17	Khan, Z.A.	11	120	1.55
18	Kim, D.	11	129	1.66
19	Pompili, D.	11	289	3.73
20	Rajasekaran, S.	11	83	1.07

Table 9.2 Subject area distribution on research in UWSN/IoUT

No.	Subject area	TNP	%
1	Computer Science	763	44.96
2	Engineering	547	32.23
3	Mathematics	148	8.72
4	Physics and Astronomy	92	5.42
5	Earth and planetary sciences	51	3.01
6	Materials Science	41	2.42
7	Energy	26	1.53
8	Decision sciences	9	0.53
9	Environmental Science	9	0.53
10	Multidisciplinary	7	0.41
11	Chemistry	4	0.24

citation data. Based on this Scopus search criteria, the following tags were retrieved: conference papers (613), articles (331), conference review (48), book chapter (14), review (13), editorial (2) book (2), letter (1) and short survey (1). The document types are as shown in Table 9.1 with the associated total number of publications (TNP) and the total number of citations (TNC). This research also discovered that a total of 1018 articles were published while 7 were still in press.

9.4   Bibliometric Analysis Results

This section provides valuable technical insights in connection with the results obtained on the distribution of UWSN/IoUTs research knowledge base both globally and in Africa. The results are presented in the subsequent subsections.

9.4.1   Global Perspective

Research Trend

Figure 9.7 shows the total number of publications (TNP) and citations (TNC) in Scopus. There is a gradual steady growth of UWSN/IoUT research from 2008, peaking in 2016 with a total publication of 111. A gradual decline is observed from 2017 up until July 2019. However, TNP and TNC counts could rise between August and December 2019.

The peak TNC at 1958 occurred in 2010. Afterwards, TNC declines steadily from 2010 onward, though an upward trend is briefly seen in 2015. From the TNP and TNC analysis, it was observed that a total of 1025 publications with a corresponding TNC count of 11,698 were recorded within the period of 2008 and July 2019. Exponential growth in TNC counts with respect to TNP was also observed. However, it must be mentioned that there has been a steady decline in both TNP and TNC counts over the past 2 years (2017–2019).

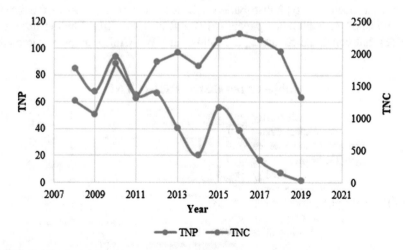

Fig. 9.7 Publication and citation trend in Scopus

Top 20 Most Productive and Cited Authors

The most influential authors are displayed in Table 9.1. They are ranked in terms of the TNP. Cui, J.-H. is the most productive with 52 TNP and, at the same time, the most cited with TNC counts of 2187. Vieira, L.F.M. is ranked second with 28 TNP and TNC counts of 708. The remaining third to twentieth ranked authors have TNP ranging between 19 and 11, respectively. It is worthy to note that author Zhou, Z. is ranked 14th, but overall second in terms of TNC counts of 960. This research noticed that most of these top-ranked authors are research collaborators or belonging to the same research niche or affiliations.

Subject Area Analysis

The top 12 subject areas in the Scopus repository are listed in Table 9.2 and depicted in Fig. 9.8. Computer Science is sitting in the first position with TNP of 763, and followed by Engineering with TNP of 547, with Mathematics having a distant third with TNP of 148. Apart from publications in Engineering and Computer Science, there has been a lot of work in other disciplines due to the multidisciplinary nature of research in the domain of UWSN/IoUT.

Top Document Types and Source Analysis

Tables 9.3 and 9.4 show the document types and document sources. While Fig. 9.9 depicts the document type distribution. For document types, most of the articles are published in conference papers (613) and articles (331). For the document sources that constitute frequent publications in UWSN/IoUT domain, conference

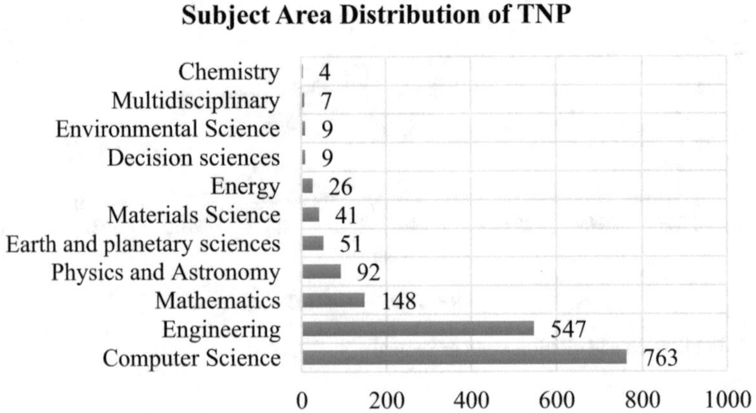

Subject Area Distribution of TNP

Fig. 9.8 Subject area distribution

Table 9.3 Document type distribution for research in UWSN/IoUT

No.	Document type	TNP	%
1	Conference paper	613	59.80
2	Article	331	32.29
3	Conference review	48	4.68
4	Book chapter	14	1.37
5	Review	13	1.27
6	Editorial	2	0.20
7	Book	2	0.20
8	Letter	1	0.10
9	Short survey	1	0.10
	Total	**1025**	**100.00**

Table 9.4 Document sources containing research in UWSN/IoUT

No.	Source types	TNP	%
1	Conference proceedings	561	54.73
2	Journals	352	34.34
3	Book Series	94	9.17
4	Books	13	1.27
5	Trade/specific Publications	5	0.49
	Total	**1025**	**100.00**

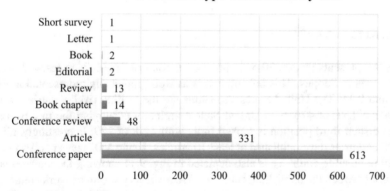

Fig. 9.9 Document type distribution

proceedings and journal are the leading publishers with TNP of 561 and 352 publications, respectively. This is in line with the logical reason that conference proceedings are a forum for researchers to put forward and discuss ideas, while journal provides the platform for more technical details on these ideas. Therefore, more publications are generally observed in conference proceedings in comparison to journals or any other document sources. Other document sources are also contributing to the overall TNP, such as book chapters, book and special issues.

Table 9.5 Top 20 countries publishing in UWSN/IoUT

No.	Country	TNP	%
1	USA	240	21.72
2	China	236	21.36
3	India	129	11.67
4	South Korea	77	6.97
5	Canada	66	5.97
6	Italy	49	4.43
7	UK	44	3.98
8	Taiwan	37	3.35
9	Brazil	34	3.08
10	Pakistan	29	2.62
11	Malaysia	20	1.81
12	Saudi Arabia	20	1.81
13	Spain	20	1.81
14	Japan	18	1.63
15	France	17	1.54
16	Iran	16	1.45
17	Norway	16	1.45
18	UAE	16	1.45
19	Australia	11	1.00
20	Bangladesh	10	0.90

Topmost Countries

Table 9.5 presents the top 20 most productive countries publishing research works in this area in Scopus. This information is as well graphically represented in Figs. 9.10 and 9.11. The United States and China are the leading rival countries, with a combined TNP of 476, with United State marginally occupying the first position. In the distant third position is ranked India with TNP of 129. Interestingly, all the top 20 countries have published at least 10 articles. Using VOSviewer software, the top 20 publishing countries is also depicted in Fig. 9.11 showing the inter-research network between the publishing countries. We see a very strong network connection between the top three productive countries, namely, USA, China and India. This is an indication of strong research collaboration between these countries.

Top Research Institutions/Affiliations

The top 20 research institutions or affiliations are outlined in Table 9.6. The University of Connecticut in the USA is ranked first, with a TNP of 69 articles and TNC of 2348, making this institution the most influential in UWSN/IoUT research. This research also noticed that most of the productive and influential authors are affiliated to the University of Connecticut. In second and third positions are the

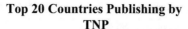

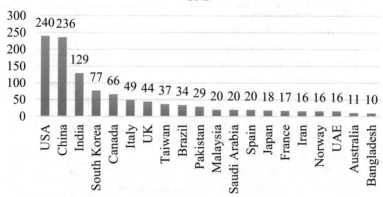

Fig. 9.10 Top 20 publishing countries

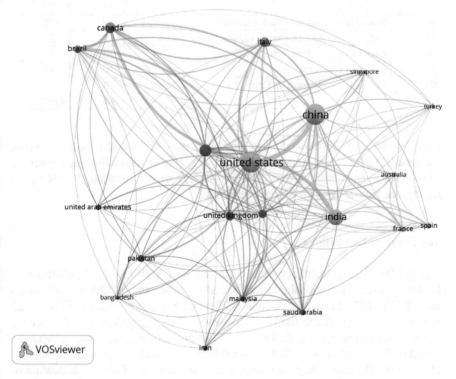

Fig. 9.11 Top 20 countries publishing

Universidade Federal de Minas Gerais (TNP = 27 and TNC = 80) and University of Ottawa, Canada (TNP = 21 and TNC =123). Interestingly, Northeastern University

Table 9.6 Top 20 Affiliation/Institutions publishing in UWSN/IoUT

No.	Affiliation/Institution by publications	TNP	TNC
1	University of Connecticut	69	2348
2	Universidade Federal de Minas Gerais	27	80
3	University of Ottawa, Canada	21	123
4	Harbin Engineering University	20	90
5	Chinese Academy of Sciences	20	162
6	Ocean University of China	20	177
7	Kyungpook National University	19	176
8	Università degli Studi di Roma La Sapienza	18	0
9	COMSATS Institute of Information Technology	18	62
10	Northeastern University	15	760
11	Glasgow Caledonian University	14	96
12	Nanjing University of Post and TeleCommunications	14	0
13	University of Alberta	14	120
14	Dalian University of Technology	13	136
15	Tianjin University	13	52
16	Memorial University of Newfoundland	12	122
17	Rutgers, The State University of New Jersey	11	123
18	NATO Undersea Research Centre	10	38
19	Kookmin University	10	17
20	Universitat Politècnica de Catalunya	10	8
Total		**368**	**4,690**

ranked tenth with TNP = 15, has citation counts of 760. On the other hand, Università Degli Studi di Roma La Sapienza with TNP = 18 and Nanjing University of Post and Telecommunications (TNP = 14) each have no citation. All the top 20 institutions have at least 10 articles published.

Top Funding Institutions

The top 20 funding institutions ranked in terms of TNP are listed in Table 9.7. In first position is National Natural Science Foundation of China (TNP = 73 and TNC = 152). In second position is National Science Foundation (TNP = 22 and TNC = 238). Interestingly, Fundamental Research Funds for the Central Universities ranked third with TNP = 14 but no citation. Additionally, there are a total of four institutions, namely, Natural Science Foundation of Jiangsu Province (TNP = 6), Doctoral Program Foundation of Institutions of Higher Education of China (TNP = 5), Fundação de Amparo à Pesquisa do Estado de Minas Gerais (TNP = 5) and Canada Research Chairs (TNP = 3) with no citations. Occupying fourth position is Office of Naval Research with TNP = 14 and TNC = 129. Worthy of note is the National Basic Research Program of China (973 Program)

Table 9.7 Top 20 Funding Institutions publishing in UWSN/IoUT

No.	Funding body by publications	TNP	TNC
1	National Natural Science Foundation of China	73	152
2	National Science Foundation	22	238
3	Fundamental Research Funds for the Central Universities	14	0
4	Office of Naval Research	14	129
5	National Research Foundation of Korea	13	3
6	China Postdoctoral Science Foundation	11	53
7	Coordenação de Aperfeiçoamento de Pessoal de Nível Superior	9	67
8	Conselho Nacional de Desenvolvimento Científico e Tecnológico	8	6
9	National Basic Research Program of China (973 Program)	7	135
10	Natural Sciences and Engineering Research Council of Canada	7	38
11	Ministry of Education	6	4
12	Natural Science Foundation of Jiangsu Province	6	0
13	Doctoral Program Foundation of Institutions of Higher Education of China	5	0
14	Fundação de Amparo à Pesquisa do Estado de Minas Gerais	5	0
15	Ministry of Education, Science and Technology	5	5
16	Ministry of Science ICT and Future Planning	5	6
17	China Scholarship Council	4	100
18	Engineering and Physical Sciences Research Council	4	39
19	National Aerospace Science Foundation of China	4	18
20	Canada Research Chairs	3	0

with TNP = 7 and China Scholarship Council with TNP = 4, having citation counts of 135 and 100, respectively.

Top 20 Keywords

The most popular and most used top 20 keywords by authors in their articles are listed in Table 9.8. This research employed VOSviewer visualization software to depict a visual representation of the top 20 most used keywords in Fig. 9.12. Underwater Sensor Networks, Sensor Networks and Underwater Acoustic are the leading top three keywords.

Top 50 Most Significant Articles/Authors

Table 9.9 lists the most cited papers, authors, year of publication, title and title of publishing source. The review paper by Heidemann et al., published in 2012 [8], is the most cited with 405 citation counts. This was closely followed by [73] with a citation count of 401 that proposed a reliable transmission routing protocol for UWSN, which is the only recent paper in the past 3 years with a high citation count. Noticeably, there are a total of 24 papers with citation counts of above 100.

Table 9.8 Top 20 keywords in UWSN/IoUT publishing

No.	Keywords	TNP	%
1	Underwater Sensor Networks	828	19.84
2	Sensor Networks	563	13.49
3	Underwater Acoustics	412	9.87
4	Sensor Nodes	345	8.27
5	Wireless Sensor Networks	267	6.40
6	Energy Efficiency	184	4.41
7	Energy Utilization	170	4.07
8	Underwater Wireless Sensor Networks	157	3.76
9	Routing Protocols	147	3.52
10	Telecommunication Networks	137	3.28
11	Underwater Acoustic Sensor Networks	116	2.78
12	Underwater Environments	108	2.59
13	Sensors	105	2.52
14	Power Management (telecommunication)	99	2.37
15	Medium Access Control	95	2.28
16	Network Routing	93	2.23
17	Underwater Sensor Network	92	2.20
18	Autonomous Underwater Vehicles	89	2.13
19	Acoustic Communications	87	2.08
20	Propagation Delays	80	1.92

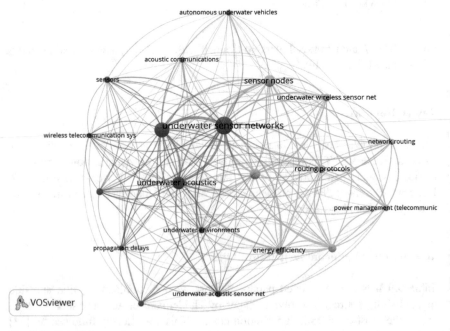

Fig. 9.12 Top 20 keywords

Table 9.9 Top 50 highly cited articles in UWSN/IoUT in Scopus

No.	Author	Title	Source title	TNC
1	Heidemann et al. [8]	Underwater sensor networks: Applications, advances and challenges	Philosophical Transactions of the Royal Society A: Mathematical, Physical and Engineering Sciences	405
2	Shen et al. [73]	A novel routing protocol providing good transmission reliability in underwater sensor networks	Journal of Internet Technology	401
3	Yan et al. [77]	DBR: Depth-based routing for underwater sensor networks	Lecture Notes in Computer Science (including subseries Lecture Notes in Artificial Intelligence and Lecture Notes in Bioinformatics)	310
4	Liu et al. [78]	Prospects and problems of wireless communication for underwater sensor networks	Wireless Communications and Mobile Computing	282
5	Tan et al. [79]	A survey of techniques and challenges in underwater localization	Ocean Engineering	244
6	Ayaz et al. [80]	A survey on routing techniques in underwater wireless sensor networks	Journal of Network and Computer Applications	179
7	Lee et al. [81]	Pressure routing for underwater sensor networks	Proceedings—IEEE INFOCOM	150
8	Zhou et al. [82]	Scalable localization with mobility prediction for underwater sensor networks	IEEE Transactions on Mobile Computing	148
9	Xie et al. [83]	Aqua-sim: An NS-2 based simulator for underwater sensor networks	MTS/IEEE Biloxi—Marine Technology for Our Future: Global and Local Challenges, OCEANS 2009	148
10	Zorzi et al. [84]	Energy-efficient routing schemes for underwater acoustic networks	IEEE Journal on Selected Areas in Communications	142
11	Noh et al. [85]	VAPR: Void-aware pressure routing for underwater sensor networks	IEEE Transactions on Mobile Computing	137
12	Arnon [86]	Underwater optical wireless communication network	Optical Engineering	128
13	Che et al. [87]	Re-evaluation of RF electromagnetic communication in underwater sensor networks	IEEE Communications Magazine	128
14	Teymorian et al. [88]	3D underwater sensor network localization	IEEE Transactions on Mobile Computing	128

(continued)

Table 9.9 (continued)

No.	Author	Title	Source title	TNC
15	Arnon & Kedar [89]	Non-line-of-sight underwater optical wireless communication network	Journal of the Optical Society of America A: Optics and Image Science, and Vision	128
16	Fazel et al. [90]	Random access compressed sensing for energy-efficient underwater sensor networks	IEEE Journal on Selected Areas in Communications	121
17	Pompili et al. [91]	Three-dimensional and two-dimensional deployment analysis for underwater acoustic sensor networks	Ad Hoc Networks	120
18	Hu & Fei [92]	QELAR: A Machine-Learning-Based Adaptive Routing Protocol for Energy-Efficient and Lifetime-Extended Underwater Sensor Networks	IEEE Transactions on Mobile Computing	116
19	Cheng et al. [93]	Underwater localization in sparse 3D acoustic sensor networks	Proceedings—IEEE INFOCOM	
20	Xiao et al. [94]	Tight performance bounds of multihop fair access for MAC protocols in wireless sensor networks and underwater sensor networks	IEEE Transactions on Mobile Computing	114
21	Zhou et al. [95]	Efficient localization for large-scale underwater sensor networks	Ad Hoc Networks	113
22	Ayaz & Abdullah [96]	Hop-by-hop dynamic addressing based (H2-DAB) routing protocol for underwater wireless sensor networks	2009 International Conference on Information and Multimedia Technology, ICIMT 2009	109
23	Guo et al. [97]	Design of a propagation-delay-tolerant MAC protocol for underwater acoustic sensor networks	IEEE Journal of Oceanic Engineering	106
24	Hsu et al. [98]	ST-MAC: Spatial-temporal MAC scheduling for underwater sensor networks	Proceedings—IEEE INFOCOM	102
25	Ammari & Das [99]	A study of k-coverage and measures of connectivity in 3D wireless sensor networks	IEEE Transactions on Computers	99
26	Hollinger et al. [22]	Underwater data collection using robotic sensor networks	IEEE Journal on Selected Areas in Communications	94
27	Felemban et al. [20]	Underwater Sensor Network Applications: A Comprehensive Survey	International Journal of Distributed Sensor Networks	81

28	Pompili et al. [100]	Distributed routing algorithms for underwater acoustic sensor networks	IEEE Transactions on Wireless Communications	80
29	Coutinho et al. [101]	Geographic and opportunistic routing for underwater sensor networks	IEEE Transactions on Computers	79
30	Zhou et al. [102]	Scalable localization with mobility prediction for underwater sensor networks	Proceedings—IEEE INFOCOM	79
31	Chen et al. [103]	Mobicast routing protocol for underwater sensor networks	IEEE Sensors Journal	78
32	Huang et al. [104]	Target tracking based on a distributed particle filter in underwater sensor networks	Wireless Communications and Mobile Computing	76
33	Hwang & Kim [105]	DFR: Directional flooding-based routing protocol for underwater sensor networks	OCEAN 2008	75
34	Domingo [62]	An overview of the internet of underwater things	Journal of Network and Computer Applications	68
35	Zhou et al. [106]	Efficient multipath communication for time-critical applications in underwater acoustic sensor networks	IEEE/ACM Transactions on Networking	65
36	Erol et al. [107]	Multi-stage underwater sensor localization using mobile beacons	Proceedings—Second Int. Conf. Sensor Technol. Appl.. SENSORCOMM 2008, Includes MESH 2008 Conf. Mesh Networks; ENOPT 2008 Energy Optim. Wireless Sensors Networks, UNWAT 2008 Under Water Sensors Systems	65
37	Guo et al. [108]	Efficient error recovery with network coding in underwater sensor networks	Ad Hoc Networks	64
38	Vajapeyam et al. [109]	Distributed space-time cooperative schemes for underwater acoustic communications	IEEE Journal of Oceanic Engineering	63
39	Liu et al. [110]	Mobi-sync: Efficient time synchronization for mobile underwater sensor networks	IEEE Transactions on Parallel and Distributed Systems	61
40	Noh et al. [111]	HydroCast: Pressure routing for underwater sensor networks	IEEE Transactions on Vehicular Technology	60

(continued)

Table 9.9 (continued)

No.	Author	Title	Source title	TNC
41	Liu et al. [112]	A Joint Time Synchronization and Localization Design for Mobile Underwater Sensor Networks	IEEE Transactions on Mobile Computing	57
42	Petrioli et al. [30]	The SUNSET framework for simulation, emulation and at-sea testing of underwater wireless sensor networks	Ad Hoc Networks	54
43	Coutinho et al. [113]	GEDAR: Geographic and opportunistic routing protocol with Depth Adjustment for mobile underwater sensor networks	2014 IEEE International Conference on Communications, ICC 2014	53
44	Noh et al. [114]	DOTS: A propagation Delay-aware Opportunistic MAC protocol for underwater sensor networks	Proceedings—International Conference on Network Protocols, ICNP	51
45	Petrioli et al. [115]	A comparative performance evaluation of MAC protocols for underwater sensor networks	OCEANS 2008	51
46	Domingo [116]	Securing underwater wireless communication networks	IEEE Wireless Communications	50
47	Chirdchoo et al. [117]	Sector-based routing with destination location prediction for underwater mobile networks	Proceedings—International Conference on Advanced Information Networking and Applications, AINA	50
48	Yu et al. [118]	WDFAD-DBR: Weighting depth and forwarding area division DBR routing protocol for UASNs	Ad Hoc Networks	49
49	Liu et al. [119]	Asymmetrical round trip based synchronization-free localization in large-scale underwater sensor networks	IEEE Transactions on Wireless Communications	49
50	Pompili & Akyildiz [120]	A multimedia cross-layer protocol for underwater acoustic sensor networks	IEEE Transactions on Wireless Communications	49

The remaining 26 papers have citation counts less than 100 and equal to 49. The most cited papers are mostly focused on overview, research challenges, practical applications, as well as research centred on routing protocols and localization.

9.5   African Perspective

Research in UWSN/IoUT is still in its infancy in Africa, with Morocco and Egypt being the only countries represented in the Scopus database. This research noticed that the earliest publication date for the two countries was in 2016. A combined total of 12 articles and 34 total citations were discovered in this search. However, it must be pointed out that there could be numerous African researchers conducting and publishing research in UWSN/IoUT in other Universities and research institutions around the globe. This can be true for those pursuing PhD degrees and engaging in research collaborations abroad. Therefore, the analysis conducted in this research may not be an exhaustive representation of research activities of African researchers. Nevertheless, based on the obtained results in this research, the most active authors are in research institutions in Morocco and Egypt. Morocco has a TNP = 7, while Egypt has a TNP = 5. The top three most cited papers are all from Morocco, namely, [74] (15), [75] (7) and [76] (5), with publication spread covering IEEE Journal of Oceanic Engineering and two editions of the International Wireless Communications and Mobile Computing Conference. The aforementioned information is contained in Table 9.10.

9.6   Discussion and Analysis

This concluding section of this research offers interpretation and implications of key findings and highlights the limitations of this study.

This research conducts bibliometric analysis and employs science mapping as a means of documenting and investigating the knowledge base accumulated in UWSN/IoUT research over the past decade. Using bibliometric analysis, this research analysed 1025 documents published in Scopus data repository between 2008 and July 2019. Globally, the papers by [8, 73] remain the most influential and cited authors in Scopus. Computer Science and Engineering are the main subject area for research in UWSN. The United States and China are the leading countries conducting research in this domain. The University of Connecticut is the leading University/Institution with most publications and citations. National Natural Science Foundation of China is the lead funding body in terms of publications. Conference and Journal are the leading document type and sources targeted by researchers in UWSN.IoUT for publication. Overall, Cui, J.-H. and his research collaborators remain the most influential author publishing in Scopus.

Table 9.10 African countries publishing in UWSN/IoUT in Scopus

No.	Author	Title	Source title	Country	TNC
1	Lmai et al. [74]	Throughput-Efficient Super-TDMA MAC Transmission Schedules in Ad Hoc Linear Underwater Acoustic Networks	IEEE Journal of Oceanic Engineering	Morocco	15
2	Jouhari et al. [76]	New greedy forwarding strategy for UWSNs geographic routing protocols	2016 International Wireless Communications and Mobile Computing Conference, IWCMC 2016	Morocco	7
3	Jouhari et al. [75]	Topology control through depth adjustment and transmission power control for UWSN routing protocols	International Conference on Wireless Networks and Mobile Communications, WINCOM 2015	Morocco	5
4	Ashri et al. [121]	A novel fractional Fourier transform-based ASK-OFDM system for underwater acoustic communications	Applied Sciences (Switzerland)	Egypt	3
5	Ashri et al. [122]	BER of FRFT-based OFDM system for underwater wireless communication	National Radio Science Conference, NRSC, Proceedings	Egypt	2
6	Jouhari et al. [123]	Implementation of bit error rate model of 16-QAM in aqua-sim simulator for underwater sensor networks	Lecture Notes in Electrical Engineering	Morocco	1
7	Alhumyani et al. [124]	Heuristic approaches for underwater sensing and processing deployment	2015 11th International Computer Engineering Conference: Today Information Society What's Next?, ICENCO 2015	Egypt	1
8	Jouhari et al. [125]	Signaling game approach to improve the MAC protocol in the underwater wireless sensor networks	International Journal of Communication Systems	Morocco	0
9	Krishnaraj et al. [126]	Deep learning model for real-time image compression in Internet of Underwater Things (IoUT)	Journal of Real-Time Image Processing	Egypt	0
10	Bennouri et al. [127]	A pursuit learning solution to underwater communications with limited mobility agents	Proceedings of the 2018 Research in Adaptive and Convergent Systems, RACS 2018	Morocco	0
11	Ammar et al. [128]	MAC Protocol-Based Depth Adjustment and Splitting Mechanism for Underwater Sensor Network (UWSN)	2018 IEEE Global Communications Conference, GLOBECOM 2018—Proceedings	Morocco	0
12	Khater et al. [129]	Contention-based MAC protocol in UWSNs: Slotted-CS-ALOHA proposed protocol	Proceedings of 2016 11th International Conference on Computer Engineering and Systems, ICCES 2016	Egypt	0
	Total citations				**34**

In Africa, research in UWSN/IoUT is still at its early stage. On the other hand, in the African region, Morocco and Egypt are the only countries in the region conducting research in this domain with a total of 13 publications and 34 citations so far.

9.6.1 Implication of Findings and Vital Missing Links

One of the vital missing links is that the UWSN/IoUT projects discussed in Sect. 9.2.2 of this chapter do not translate or result into impactful R&D outputs (publications, research projects, local innovations, indigenous solutions) as reported and analysed in Sect. 9.5 of this chapter. Apart from the UWSN/IoUT projects discussed in Sect. 9.2.2, there are many African-based foreign companies specializing in selling smart underwater technologies and offering consultancy services to interested local consumers. However, the findings of this research have obviously revealed that indigenous institutions (universities, R&D institutes, and innovation hubs), local experts and local products/resources were not extensively employed for these UWSN/IoUT projects. The implication of this is that these UWSN/IoUT projects are more or less cases of technology transfer (or turnkey projects) and not genuine knowledge or know-how transfer. This limited patronage of local experts and indigenous institutions have slowed down the realization of visible technological impact and tangible R&D outputs from UWSN/IoUT projects in Africa.

Other vital missing links for UWSN/IoUT projects in Africa are, namely, low/restricted access to current researches (from leading databases and online libraries) and technological trends in UWSN/IoUT research area; dearth of long-term sponsors and funding for UWSN/IoUT research projects; high cost of backbone infrastructural building blocks; and limited public-private partnerships for collaborative research among corporate bodies, government agencies, universities, research institutes, innovation hubs and R&D centres locally and regionally.

9.6.2 Research Directions and Recommendations

Useful pointers for facilitating the successful actualization of UWSN/IoUT research projects in Africa are, namely, establishing strong and cohesive linkages between local and regional stakeholders (firms, agencies, universities, R&D institutes, innovation hubs) for effective public-private partnerships; subscription and opening up access to up-to-date research works and cutting-edge technologies from leading databases and information sources; promotion of indigenous innovations and more patronage of local experts; and attracting committed sponsors, concretizing funding plans and investing smartly in core infrastructure.

Limitations of this study which will be addressed in an upcoming research work are, namely, (1) scope of years under examination/analysis, (2) focusing analysis on Scopus database (this choice of Scopus database has been earlier argued that it captures a substantial coverage of research work in Engineering and Physical Sciences which UWSN/IoUT falls into, and hence giving a sufficient representation of research trends in this domain), (3) Omission of non-English journals may not capture countries publishing in other languages such as Russia that might be conducting lots of research in UWSN/IoUT. While National Natural Science Foundation of China is the leading funding body in terms of publication, there are still many important unanswered questions because this performance indicator cannot be directly linked to or reliably used as a true picture of pertinent R&D funding and infrastructure at the disposal of these funding bodies.

9.7 Conclusion

Advancements in CC and IoT have revolutionized TWSN communication which has paved the way for the practical design of UWSN in order to support the emergence of IoUT. UWSN is an essential asset for smart cities as about 70% of the earth is covered by water bodies, and it is evident that oceans largely regulate global climatic conditions. The need for better environmental monitoring within the context of smart cities and the recent spate of global natural disasters has further aroused research interest in IoUT. This is motivated by a number of UWSN innovative solutions and smart underwater technologies. While these inventions hold promising prospects for technologically advanced countries, the same assertion cannot be made for most African countries due to challenges inherent in R&D activities into critical IoUT/UWSN projects in the region. This chapter conducts a bibliometric analysis that systematically highlights the knowledge base for core research works in UWSN globally and within the African region. This work employs bibliometric analysis and VOSviewer science mapping software tool to analyse 1025 scientific research publications in SCOPUS-indexed database over a period of 10 years (2008 to 2019) in the field of UWSN. This research methodically investigates and interprets the ensuing findings of the analysis in order to carve out useful technical approaches and important procedures for exploiting the knowledge domain and intellectual structure of UWSN research within the African context. Furthermore, this analysis identifies and highlights vital missing links, essential research directions and unique technical contributions and recommendations that will be of relevance in facilitating the successful actualization of IoUT/UWSN research projects in Africa.

Acknowledgement The authors would like to thank Ahmadu Bello University, Zaria, Nigeria, and University of Johannesburg, South Africa, for their support and affording the resources to complete this research work.

References

1. I.F. Akyildiz, D. Pompili, T. Melodia, Underwater acoustic sensor networks: Research challenges. Ad Hoc Netw. **3**(3), 257–279 (2005). https://doi.org/10.1016/j.adhoc.2005.01.004
2. F. Al-Turjman, S. Alturjman, Confidential smart-sensing framework in the IoT era. J. Supercomput. **74**(10), 5187–5198 (2018). https://doi.org/10.1007/s11227-018-2524-1
3. U.D. Ulusar, F. Al-Turjman, G. Celik. *An Overview of Internet of Things and Wireless Communications*. International Conference on Computer Science and Engineering (UBMK), Antalya, Turkey (2017), pp. 506–509. doi: https://doi.org/10.1109/UBMK.2017.8093446
4. S. Mahmoudzadeh, D.M.W. Powers, A. Atyabi, UUV's hierarchical DE-based motion planning in a semi dynamic underwater wireless sensor network. IEEE Trans. Cyber. **49**(8), 2992–3005 (2019). https://doi.org/10.1109/TCYB.2018.2837134
5. N. Saeed, M.-S. Alouini, T. Y. Al-Naffouri. *To Boldly go Where no Sensor has Gone Before: The Movement to Place IoT in Radical New Spaces* (2019). https://www.comsoc.org/publications/ctn/boldly-go-where-no-sensor-has-gone-movement-place-iot-radical-new-spaces. Accessed 30 June 2019
6. A.J. Watt, M.R. Phillips, C.E. Campbell, I. Wells, S. Hole, Wireless sensor networks for monitoring underwater sediment transport. Sci. Total Environ. **667**, 160–165 (2019). https://doi.org/10.1016/j.scitotenv.2019.02.369
7. S.M. Demir, F. Al-Turjman, A. Muhtaroglu, Energy scavenging methods for WBAN applications: A review. IEEE Sensors J. **18**(16), 6477–6488 (2018). https://doi.org/10.1109/JSEN.2018.2851187
8. J. Heidemann, M. Stojanovic, M. Zorzi, Underwater sensor networks: Applications, advances and challenges. Philos. Trans. R. Soc. A Math. Phys. Eng. Sci. **370**(1958), 158–175 (2012). https://doi.org/10.1098/rsta.2011.0214
9. F.H.M.B. Lima, L.F.M. Vieira, A.B. Vieira, M.A.M. Vieira, J.A.M. Nacif, Water ping: ICMP for the Internet of underwater things. Comput. Netw. **152**, 54–63 (2019). https://doi.org/10.1016/j.comnet.2019.01.009
10. E. Liou, C. Kao, C. Chang, Y. Lin, C. Huang. *Internet of Underwater Things: Challenges and Routing Protocols*. 4th IEEE International Conference on Applied System Innovation (ICASI), Chiba (2018), pp. 1171–1174. doi:https://doi.org/10.1109/ICASI.2018.8394494
11. L. J. Poncha, S. Abdelhamid, S. Alturjman, E. Ever, F. Al-Turjman *5G in a Convergent Internet of Things Era: An Overview*. In IEEE International Conference on Communications Workshops (ICC Workshops), (Kansas City, MO, 2018), pp. 1–6. https://doi.org/10.1109/ICCW.2018.8403748
12. Z. Mbabela. *Age of the drone and autonomous underwater* (2018).https://news.mandela.ac.za/News/Age-of-the-Drone-and-Autonomous-Underwater-Vehicle. Accessed 19 June 2018
13. F. Al-Turjman, Information-centric sensor networks for cognitive IoT: An overview. Ann. Telecommun. **72**(1), 3–18 (2017). https://doi.org/10.1007/s12243-016-0533-8
14. K.M. Awan, P.A. Shah, K. Iqbal, S. Gillani, W. Ahmad, Y. Nam, Underwater wireless sensor networks: A review of recent issues and challenges. Wirel. Commun. Mob. Comput. **2019**, 6470359 (2019). https://doi.org/10.1155/2019/6470359
15. E. Colmenar, F. Al-Turjman, M. Biglarbegian. *Data Delivery and Gathering in IoT Applications: An Overview*. 39th Annual IEEE Conference on Local Computer Networks Workshops, (Edmonton, AB, 2014), pp. 790–795
16. G. Han, X. Long, C. Zhu, M. Guizani, Y. Bi, W. Zhang, An AUV location prediction-based data collection scheme for underwater wireless sensor networks. IEEE Trans. Veh. Technol. **68**(6), 6037–6049 (2019). https://doi.org/10.1109/TVT.2019.2911694
17. Z. Hong, X. Pan, P. Chen, X. Su, N. Wang, W. Lu, A topology control with energy balance in underwater wireless sensor networks for IoT-based application. Sensors **18**(7), 2306 (2018). https://doi.org/10.3390/s18072306
18. E.M. Dogo, A.F. Salami, C.O. Aigbavboa, T. Nkonyana, Taking cloud computing to the extreme edge: A review of mist computing for smart cities and industry 4.0 in Africa, in *Edge computing*, ed. by F. Al-Turjman, (Springer, Cham, 2019a)

19. E.M. Dogo, A.F. Salami, N.I. Nwulu, C.O. Aigbavboa, Blockchain and Internet of things-based technologies for intelligent water management system, in *Artificial intelligence in IoT*, ed. by F. Al-Turjman, (Springer, Cham, 2019b)
20. E. Felemban, F.K. Shaikh, U.M. Qureshi, A.A. Sheikh, S.B. Qaisar, Underwater sensor network applications: A comprehensive survey. Int. J. Distr. Sens. Net. (2015). https://doi.org/10.1155/2015/896832
21. Q. Guan, F. Ji, Y. Liu, H. Yu, W. Chen, Distance-vector-based opportunistic routing for underwater acoustic sensor networks. IEEE Internet Things J. **6**(2), 3831–3839 (2019). https://doi.org/10.1109/JIOT.2019.2891910
22. G.A. Hollinger, S. Choudhary, P. Qarabaqi, C. Murphy, U. Mitra, G.S. Sukhatme, et al., Underwater data collection using robotic sensor networks. IEEE J. Select. Areas Commun. **30**(5), 899–911 (2012). https://doi.org/10.1109/JSAC.2012.120606
23. A. Darehshoorzadeh, A. Boukerche, Underwater sensor networks: A new challenge for opportunistic routing protocols. IEEE Commun. Mag. **53**(11), 98–107 (2015). https://doi.org/10.1109/MCOM.2015.7321977
24. A. Davis, H. Chang, *Underwater wireless sensor networks* (IEEE OCEANS, Virginia Beach, VA, 2012), pp. 1–5
25. M. Hussaini, H. Bello-Salau, A.F. Salami, F. Anwar, A.H. Abdalla, M.R. Islam, Enhanced clustering routing protocol for power-efficient gathering in wireless sensor network. Int. J. Commun. Network. Inform. Secur. **4**(1), 18–28 (2012)
26. J. Lloret, Underwater sensor nodes and networks. Sensors **13**(9), 11782–11796 (2013). https://doi.org/10.3390/s130911782
27. H. Bello-Salau, A.J. Onumanyi, A.F. Salami, S. Muslim, W.M. Audu, U. Abdullahi, Improved clustering routing protocol for low-energy adaptive cluster-based routing in wireless sensor network. ATBU J. Sci. Technol. Edu. **6**(3), 113–125 (2018)
28. E. Cayirci, H. Tezcan, Y. Dogan, V. Coskun, Wireless sensor networks for underwater survelliance systems. Ad Hoc Netw. **4**(4), 431–446 (2006). https://doi.org/10.1016/j.adhoc.2004.10.008
29. G. Han, J. Jiang, N. Bao, L. Wan, M. Guizani, Routing protocols for underwater wireless sensor networks. IEEE Commun. Mag. **53**(11), 72–78 (2015). https://doi.org/10.1109/MCOM.2015.7321974
30. C. Petrioli, R. Petroccia, J.R. Potter, D. Spaccini, The SUNSET framework for simulation, emulation and at-sea testing of underwater wireless sensor networks. Ad Hoc Netw. **34**, 224–238 (2015). https://doi.org/10.1016/j.adhoc.2014.08.012
31. M. Shakir, M.A. Khan, S.A. Malik, Izhar-ul-Haq, Design of underwater sensor networks for water quality monitoring. World Appl. Sci. J. **17**(11), 1441–1444 (2012)
32. M. Waldmeyer, H.-P. Tan, W. K. G. Seah. *Multi-stage AUV-Aided Localization for Underwater Wireless Sensor Networks*. IEEE Workshops of International Conference on Advanced Information Networking and Applications, (Singapore, 2011), pp. 908–913. doi: https://doi.org/10.1109/WAINA.2011.90
33. G. Xu, W. Shen, X. Wang, Applications of wireless sensor networks in marine environment monitoring: A survey. Sensors **14**(9), 16932–16954 (2014). https://doi.org/10.3390/s140916932
34. M.C. Domingo, R. Prior, Energy analysis of routing protocols for underwater wireless sensor networks. Comput. Commun. **31**(6), 1227–1238 (2008). https://doi.org/10.1016/j.comcom.2007.11.005
35. A. Khan, L. Jenkins. *Undersea Wireless Sensor Network for Ocean Pollution Prevention*. 3rd International Conference on Communication Systems Software and Middleware and Workshops (COMSWARE '08), Bangalore (2008), pp. 2–8. doi: https://doi.org/10.1109/COMSWA.2008.4554369
36. J. Lloret, S. Sendra, M. Garcia, G. Lloret. *Group-Based Underwater Wireless Sensor Network for Marine Fish Farms*. IEEE GLOBECOM Workshops (GCWkshps), Houston, TX (2011), pp. 115–119. https://doi.org/10.1109/GLOCOMW.2011.6162361

37. A. F. Salami, F. Anwar, A. M. Aibinu, H. Bello-Salau, A. H. Abdalla. *Investigative Analysis of Clustering Routing Protocols for Scalable Sensor Networks*. 4th IEEE International Conference on Mechatronics (ICOM), (Kuala Lumpur, Malaysia, 2011a), pp. 011–015
38. A.F. Salami, H. Bello-Salau, F. Anwar, A.M. Aibinu, A novel biased energy distribution (BED) technique for cluster-based routing in wireless sensor networks. Int. J. Smart Sens. Intellig. Syst. **4**(2), 161–173 (2011b). https://doi.org/10.21307/ijssis-2017-433
39. A.F. Salami, F. Anwar, A.U. Priantoro, An investigation into clustering routing protocols for wireless sensor networks. Sens. Transd. **106**(7), 48–61 (2009)
40. S. Zhang, J. Yu, A. Zhang, L. Yang, Y. Shu, Marine vehicle sensor network architecture and protocol designs for ocean observation. Sensors **12**(1), 373–390 (2012). https://doi.org/10.3390/s120100373
41. P. Kumar, P. Kumar, P. Priyadarshini, Srija. *Underwater Acoustic Sensor Network for Early Warning Generation*. 2012 Oceans, Hampton Roads, VA (2012), pp. 1–6. https://doi.org/10.1109/OCEANS.2012.6405009
42. N. F. Henry, O. N. Henry, Wireless sensor networks based pipeline vandalisation and oil spillage monitoring and detection: Main benefits for Nigeria oil and gas sectors. SIJ Trans. Comp. Sci. Eng. Appl. **3**(1), 1–6 (2015)
43. A. F. Salami, S. M. S. Bari, F. Anwar, S. Khan. *Feasibility Analysis Of Clustering Routing Protocols For Multipurpose Sensor Networking*. 2nd International Conference on Multimedia and Computational Intelligence (ICMCI), (Shanghai, China, 2010), pp. 432–435
44. A.F. Salami, E.M. Dogo, N.I. Nwulu, B.S. Paul, Toward sustainable domestication of smart IoT mobility solutions for the visually impaired persons in Africa, in *Technological trends in improved mobility of the visually impaired*, ed. by S. Paiva, (Springer, Cham, 2020). https://doi.org/10.1007/978-3-030-16450-8_11
45. S. Srinivas, P. Ranjitha, R. Ramya, G. K Narendra. *Investigation of Oceanic Environment Using Large-Scale UWSN and UANETs*. 8th International Conference on Wireless Communications, Networking and Mobile Computing, Shanghai (2012), pp. 1–5. doi: https://doi.org/10.1109/WiCOM.2012.6478552
46. S. Tyan, S. Oh, AUV-RM: Underwater sensor network scheme for AUV based river monitoring. Res. Trend Comp. Appl. **24**, 53–55 (2013)
47. C. Alippi, R. Camplani, C. Galperti, M. Roveri, A robust, adaptive, solar-powered WSN framework for aquatic environmental monitoring. IEEE Sensors J. **11**(1), 45–55 (2011). https://doi.org/10.1109/JSEN.2010.2051539
48. A. Caiti, V. Calabrò, A. Munafò, G. Dini, A. Lo Duca, Mobile underwater sensor networks for protection and security: Field experience at the UAN11 experiment. J. Field Robot. **30**(2), 237–253 (2013). https://doi.org/10.1002/rob.21447
49. S. Kemna, M.J. Hamilton, D.T. Hughes, K.D. Lepage, Adaptive autonomous underwater vehicles for littoral surveillance. Intell. Serv. Robot. **4**, 245–258 (n.d.). https://doi.org/10.1007/s11370-011-0097-4
50. K. Casey, A. Lim, G. Gerry Dozier, A sensor network architecture for tsunami detection and response. Int. J. Distri. Sens. Netw. **4**(1), 27–42 (2008). https://doi.org/10.1080/15501320701774675
51. A. Pirisi, F. Grimaccia, M. Mussetta, R. E. Zich, R. Johnstone, M. Palaniswami, et al. *Optimization of an Energy Harvesting Buoy for Coral Reef Monitoring*. IEEE Congress on Evolutionary Computation, Cancun (2013), pp. 629–634. doi: https://doi.org/10.1109/CEC.2013.6557627
52. M. Stojanovic. *On The Relationship Between Capacity And Distance In An Underwater Acoustic Communication Channel*. Proceedings of the 1st ACM International Workshop on Underwater Networks (WUWNet '06), Los Angeles, CA, USA (2006), pp. 41–47. doi: https://doi.org/10.1145/1161039.1161049
53. D. P. Williams. *On optimal AUV Track-Spacing for Underwater Mine Detection*. IEEE International Conference on Robotics and Automation, (Anchorage, AK, 2010), pp. 4755–4762. doi: https://doi.org/10.1109/ROBOT.2010.5509435

54. S. Zhou, P. Willett, Submarine location estimation via a network of detection-only sensors. IEEE Trans. Signal Process. **55**(6), 3104–3115 (2007). https://doi.org/10.1109/TSP.2007.893970

55. E.M. Dogo, A. Salami, S. Salman, Feasibility analysis of critical factors affecting cloud computing in Nigeria. Int. J. Cloud Comp. Serv. Sci. **2**(4), 276–287 (2013). https://doi.org/10.11591/closer.v2i4.4162

56. H. Bello-Salau, A.F. Salami, F. Anwar, A.M. Aibinu. *Evaluation of Radio Propagation Techniques for Hierarchical Sensor Networks*. 4th IEEE International Conference on Mechatronics (ICOM), (Kuala Lumpur, Malaysia, 2011b), pp. 001–005

57. H. Bello-Salau, A.F. Salami, F. Anwar, M.R. Islam, Analysis of radio model performance for clustering sensor networks. Sens. Transd. **128**(5), 27–38 (2011a)

58. G. Han, A. Qian, C. Zhang, Y. Wang, J.P. Joel, C. Rodrigues, Localization algorithms in large-scale underwater acoustic sensor networks: A quantitative comparison. Int. J. Distr. Sens. Netw **10**(3), 379382 (2014). https://doi.org/10.1155/2014/379382

59. Y. Guo, Y. Liu, Localization for anchor-free underwater sensor networks. Comput. Electr. Eng. **39**(6), 1812–1821 (2013). https://doi.org/10.1016/j.compeleceng.2013.02.001

60. Z. Zhou, B. Yao, R. Xing, L. Shu, S. Bu, E-CARP: An energy efficient routing protocol for UWSNs in the Internet of underwater things. IEEE Sensors J. **16**(11), 4072–4082 (2016). https://doi.org/10.1109/JSEN.2015.2437904

61. S.Y. Chen, T.T. Juang, Y. Lin, I.C. Tsai, A low propagation delay multi-path routing protocol for underwater sensor networks. J. Int. Technol **11**, 153–165 (2010)

62. M.C. Domingo, An overview of the Internet of underwater things. J. Netw. Comput. Appl. **35**(6), 1879–1890 (2012). https://doi.org/10.1016/j.jnca.2012.07.012

63. C. Kao, Y. Lin, G. Wu, C. Huang, A comprehensive study on the Internet of underwater things: Applications, challenges, and channel models. Sensors **17**(7), 1477 (2017). https://doi.org/10.3390/s17071477

64. J.-H. Cui, J. Kong, M. Gerla, S. Zhou, The challenges of building scalable mobile underwater wireless sensor networks for aquatic applications. IEEE Netw. **20**(3), 12–18 (2006). https://doi.org/10.1109/MNET.2006.1637927

65. Shell World Nigeria. *SCiN Innovation: Remote Underwater Surveys* (2015).https://www.shell.com.ng/media/shell-world-nigeria/_jcr_content/par/textimage.stream/1480321977348/37e0787be55dbcd532a792a738f43885d57678b0/shell-world-nigeria032015-q1.pdf. Accessed 6 June 2019

66. S. Keedwell. *Advancing the Art of Subsea Inspection* (2011).https://www.offshore-mag.com/subsea/article/16755201/advancing-the-art-of-subsea-inspection. Accessed 7 June 2019

67. B.F. Ndiaye. *Climate and Via Water Cafe: Capture and Share Continuous Data on Water Quality of Niger River Around Bamako* (2018),https://www.climatescan.nl/projects/2436/detail. Accessed 28 June 2018

68. Dutch Water Sector. *Indymo Underwater Drone Collects Water Quality Data in Niger River, Mali* (2018).https://www.dutchwatersector.com/news/indymo-underwater-drone-collects-water-quality-data-in-niger-river-mali, Accessed 27 June 2018

69. I. Wangui *Sensors Boost Environmental Journalism and Citizen Engagement in Five African Countries* (2018).https://ijnet.org/en/story/sensors-boost-environmental-journalism-and-citizen-engagement-five-african-countries. Accessed 29 June 2018

70. C. Srimathi, S. Park, N. Rajesh. *Proposed Framework For Underwater Sensor Cloud For Environmental Monitoring*. 5th International Conference on Ubiquitous and Future Networks, ICUFN 2013, (Da Nang, 2013). pp. 104–109. doi: https://doi.org/10.1109/ICUFN.2013.6614788

71. P. Mongeon, A. Paul-Hus, The journal coverage of web of science and scopus: A comparative analysis. Scientometrics **106**(1), 213–228 (2016). https://doi.org/10.1007/s11192-015-1765-5

72. D. Moher, A. Liberati, J. Tetzlaff, D.G. Altman, PRISMA Group, Preferred reporting items for systematic reviews and meta-analyses: The PRISMA statement. PLoS Med. **6**(7), e1000097 (2009). https://doi.org/10.1371/journal.pmed.1000097

73. J. Shen, H. Tan, J. Wang, J. Wang, S. Lee, A novel routing protocol providing good transmission reliability in underwater sensor networks. J. Inter. Technol. **16**(1), 171–178 (2015). https://doi.org/10.6138/JIT.2014.16.1.20131203e

74. S. Lmai, M. Chitre, C. Laot, S. Houcke, Throughput-efficient super-TDMA MAC transmission schedules in ad hoc linear underwater acoustic networks. IEEE J. Ocean. Eng. **42**(1), 156–174 (2017). https://doi.org/10.1109/JOE.2016.2537659

75. M. Jouhari, K. Ibrahimi, M. Benattou. *Topology Control Through Depth Adjustment and Transmission Power Control for UWSN Routing Protocols.* International Conference on Wireless Networks and Mobile Communications, WINCOM 2015 (2016a), https://doi.org/10.1109/WINCOM.2015.7381310

76. M. Jouhari, K. Ibrahimi, M. Benattou, A..Kobbane. *New Greedy Forwarding Strategy for UWSNs Geographic Routing Protocols.* 12th IEEE International Wireless Communications and Mobile Computing Conference, IWCMC 2016 (2016b), pp. 388–393. https://doi.org/10.1109/IWCMC.2016.7577089

77. H. Yan, Z.J. Shi, J.-H. Cui, *DBR: Depth-Based Routing For Underwater Sensor Networks* (2008). https://doi.org/10.1007/978-3-540-79549-0_7

78. L. Liu, S. Zhou, J.-H. Cui, Prospects and problems of wireless communication for underwater sensor networks. Wirel. Commun. Mob. Comput. **8**(8), 977–994 (2008). https://doi.org/10.1002/wcm.654

79. H. Tan, R. Diamant, W.K.G. Seah, M. Waldmeyer, A survey of techniques and challenges in underwater localization. Ocean Eng. **38**(14–15), 1663–1676 (2011). https://doi.org/10.1016/j.oceaneng.2011.07.017

80. M. Ayaz, I. Baig, A. Abdullah, I. Faye, A survey on routing techniques in underwater wireless sensor networks. J. Netw. Comput. Appl. **34**(6), 1908–1927 (2011). https://doi.org/10.1016/j.jnca.2011.06.009

81. U. Lee, P. Wang, Y. Noh, L.F.M. Vieira, M. Gerla, J.-H. Cui. *Pressure Routing for Underwater Sensor Networks.* In IEEE INFOCOM 2010, San Diego, CA (2010). https://doi.org/10.1109/INFCOM.2010.5461986

82. Z. Zhou, Z. Peng, J.-H. Cui, Z. Shi, A. Bagtzoglou, Scalable localization with mobility prediction for underwater sensor networks. IEEE Trans. Mob. Comput. **10**(3), 335–348 (2011b). https://doi.org/10.1109/TMC.2010.158

83. P. Xie, Z. Zhou, Z. Peng, H. Yan, T. Hu, J.-H. Cui, et al.. *Aqua-sim: An NS-2 Based Simulator for Underwater Sensor Networks.* OCEANS 2009, Biloxi, MS (2009), pp. 1–7.https://doi.org/10.23919/OCEANS.2009.5422081

84. M. Zorzi, P. Casari, N. Baldo, A.F. Harris III, Energy-efficient routing schemes for underwater acoustic networks. IEEE J. Select. Areas Commun. **26**(9), 1754–1766 (2008). https://doi.org/10.1109/JSAC.2008.081214

85. Y. Noh, U. Lee, P. Wang, B.S.C. Choi, M. Gerla, VAPR: Void-aware pressure routing for underwater sensor networks. IEEE Trans. Mob. Comput. **12**(5), 895–908 (2013). https://doi.org/10.1109/TMC.2012.53

86. S. Arnon, Underwater optical wireless communication network. Opt. Eng. **49**(1) (2010). https://doi.org/10.1117/1.3280288

87. X. Che, I. Wells, G. Dickers, P. Kear, X. Gong, Re-evaluation of RF electromagnetic communication in underwater sensor networks. IEEE Commun. Mag. **48**(12), 143–151 (2010). https://doi.org/10.1109/MCOM.2010.5673085

88. A.Y. Teymorian, W. Cheng, L. Ma, X. Cheng, X. Lu, Z. Lu, 3D underwater sensor network localization. IEEE Trans. Mob. Comput. **8**(12), 1610–1621 (2009). https://doi.org/10.1109/TMC.2009.80

89. S. Arnon, D. Kedar, Non-line-of-sight underwater optical wireless communication network. J. Opt. Soc. Am. A **26**(3), 530–539 (2009). https://doi.org/10.1364/JOSAA.26.000530

90. F. Fazel, M. Fazel, M. Stojanovic, Random access compressed sensing for energy-efficient underwater sensor networks. IEEE J. Select. Areas Commun. **29**(8), 1660–1670 (2011). https://doi.org/10.1109/JSAC.2011.110915

91. D. Pompili, T. Melodia, I.F. Akyildiz, Three-dimensional and two-dimensional deployment analysis for underwater acoustic sensor networks. Ad Hoc Netw. **7**(4), 778–790 (2009). https://doi.org/10.1016/j.adhoc.2008.07.010

92. T. Hu, Y. Fei, QELAR: A machine-learning-based adaptive routing protocol for energy-efficient and lifetime-extended underwater sensor networks. IEEE Trans. Mob. Comput. **9**(6), 796–809 (2010). https://doi.org/10.1109/TMC.2010.28

93. W. Cheng, A.Y. Teymorian, L. Ma, X. Cheng, X. Lu, Z. Lu. *Underwater Localization in Sparse 3D Acoustic Sensor Networks.* In 27th IEEE Communications Society Conference on Computer Communications, Phoenix, AZ (2008). pp. 798–806. doi:https://doi.org/10.1109/INFOCOM.2007.56

94. Y. Xiao, M. Peng, J. Gibson, G.G. Xie, D. Du, A.V. Vasilakos, Tight performance bounds of multihop fair access for MAC protocols in wireless sensor networks and underwater sensor networks. IEEE Trans. Mob. Comput. **11**(10), 1538–1554 (2012). https://doi.org/10.1109/TMC.2011.190

95. Z. Zhou, J.-H. Cui, S. Zhou, Efficient localization for large-scale underwater sensor networks. Ad Hoc Netw. **8**(3), 267–279 (2010). https://doi.org/10.1016/j.adhoc.2009.08.005

96. M. Ayaz, A. Abdullah. *Hop-by-hop Dynamic Addressing Based (H2-DAB) Routing Protocol for Underwater Wireless Sensor Networks.* Paper presented at the 2009 International Conference on Information and Multimedia Technology, ICIMT 2009, (Jeju Island, 2009). pp. 436–441. https://doi.org/10.1109/ICIMT.2009.70

97. X. Guo, M.R. Frater, M.J. Ryan, Design of a propagation-delay-tolerant MAC protocol for underwater acoustic sensor networks. IEEE J. Ocean. Eng. **34**(2), 170–180 (2009a). https://doi.org/10.1109/JOE.2009.2015164

98. C. Hsu, K. Lai, C. Chou, K.C. Lin. ST-MAC: Spatial-temporal MAC scheduling for underwater sensor networks. IEEE INFOCOM 2009 (Rio de Janeiro, 2009). pp. 1827–1835. https://doi.org/10.1109/INFCOM.2009.5062103

99. H.M. Ammari, S. Das, A study of k-coverage and measures of connectivity in 3D wireless sensor networks. IEEE Trans. Comput. **59**(2), 243–257 (2010). https://doi.org/10.1109/TC.2009.166

100. D. Pompili, T. Melodia, I.F. Akyildiz, Distributed routing algorithms for underwater acoustic sensor networks. IEEE Trans. Wirel. Commun. **9**(9), 2934–2944 (2010). https://doi.org/10.1109/TWC.2010.070910.100145

101. R.W.L. Coutinho, A. Boukerche, L.F.M. Vieira, A.A.F. Loureiro, Geographic and opportunistic routing for underwater sensor networks. IEEE Trans. Comput. **65**(2), 548–561 (2016). https://doi.org/10.1109/TC.2015.2423677

102. Z. Zhou, J.-H. Cui, A. Bagtzoglou. *Scalable Localization With Mobility Prediction For Underwater Sensor Networks.* 27th IEEE Communications Society Conference on Computer Communications, (Phoenix, AZ, 2008). pp. 211–215. doi: https://doi.org/10.1109/INFOCOM.2007.287

103. Y. Chen, Y. Lin, Mobicast routing protocol for underwater sensor networks. IEEE Sensors J. **13**(2), 737–749 (2013). https://doi.org/10.1109/JSEN.2012.2226877

104. Y. Huang, W. Liang, H. Yu, Y. Xiao, Target tracking based on a distributed particle filter in underwater sensor networks. Wirel. Commun. Mob. Comput. **8**(8), 1023–1033 (2008). https://doi.org/10.1002/wcm.660

105. D. Hwang, D. Kim. *DFR: Directional flooding-based routing protocol for underwater sensor networks.* Paper presented at the Oceans 2008 (Quebec City, QC, 2008). https://doi.org/10.1109/OCEANS.2008.5151939

106. Z. Zhou, Z. Peng, J.-H. Cui, Z. Shi, Efficient multipath communication for time-critical applications in underwater acoustic sensor networks. IEEE/ACM Trans. Networking **19**(1), 28–41 (2011a). https://doi.org/10.1109/TNET.2010.2055886

107. M. Erol, L.F.M. Vieira, A. Caruso, F. Paparella, M. Gerla, S. Oktug. *Multistage Underwater Sensor Localization Using Mobile Beacons.* In Second International Conference on Sensor Technologies and Applications (sensorcomm 2008) (Cap Esterel, 2008), pp. 710–714. doi: https://doi.org/10.1109/SENSORCOMM.2008.32

108. Z. Guo, B. Wang, P. Xie, W. Zeng, J.-H. Cui, Efficient error recovery with network coding in underwater sensor networks. Ad Hoc Netw. **7**(4), 791–802 (2009b). https://doi.org/10.1016/j.adhoc.2008.07.011

109. M. Vajapeyam, S. Vedantam, U. Mitra, J.C. Preisig, M. Stojanovic, Distributed space-time cooperative schemes for underwater acoustic communications. IEEE J. Ocean. Eng. **33**(4), 489–501 (2008). https://doi.org/10.1109/JOE.2008.2005338

110. J. Liu, Z. Zhou, Z. Peng, J.-H. Cui, M. Zuba, L. Fiondella, Mobi-sync: Efficient time synchronization for mobile underwater sensor networks. IEEE Trans. Parall Distri. Syst. **24**(2), 406–416 (2013). https://doi.org/10.1109/TPDS.2012.71

111. Y. Noh, U. Lee, S. Lee, P. Wang, L.F.M. Vieira, J.-H. Cui, et al., HydroCast: Pressure routing for underwater sensor networks. IEEE Trans. Veh. Technol. **65**(1), 333–347 (2016). https://doi.org/10.1109/TVT.2015.2395434

112. J. Liu, Z. Wang, J.-H. Cui, S. Zhou, B. Yang, A joint time synchronization and localization design for mobile underwater sensor networks. IEEE Trans. Mob. Comput. **15**(3), 530–543 (2016). https://doi.org/10.1109/TMC.2015.2410777

113. R.W.L. Coutinho, A. Boukerche, L.F.M. Vieira, A.A.F. Loureiro. *GEDAR: Geographic and Opportunistic Routing Protocol with Depth Adjustment for Mobile Underwater Sensor Networks*. 1st IEEE International Conference on Communications, (ICC 2014) (Sydney, NSW, 2014). pp. 251–256. doi:https://doi.org/10.1109/ICC.2014.6883327

114. Y. Noh, P. Wang, U. Lee, D. Torres, M. Gerla. *Dots: A Propagation Delay-Aware Opportunistic Mac Protocol For Underwater Sensor Networks*. International Conference on Network Protocols, ICNP; 18th IEEE International Conference on Network Protocols, ICNP'10, Kyoto (2010). pp. 183–192. https://doi.org/10.1109/ICNP.2010.5762767

115. C. Petrioli, R. Petroccia, M. Stojanovic. *A Comparative Performance Evaluation of MAC Protocols for Underwater Sensor Networks*. Oceans 2008; Oceans 2008, Quebec City, QC (2008). https://doi.org/10.1109/OCEANS.2008.5152042

116. M. Domingo, Securing underwater wireless communication networks. IEEE Wirel. Commun. **18**(1), 22–28 (2011). https://doi.org/10.1109/MWC.2011.5714022

117. N. Chirdchoo, W. Soh, K.C. Chua. *Sector-Based Routing with Destination Location Prediction for Underwatermobile Networks*. International Conference on Advanced Information Networking and Applications Workshops, WAINA 2009, (Bradford, 2009). pp. 1148–1153. doi:https://doi.org/10.1109/WAINA.2009.105

118. H. Yu, N. Yao, T. Wang, G. Li, Z. Gao, G. Tan, WDFAD-DBR: Weighting depth and forwarding area division DBR routing protocol for UASNs. Ad Hoc Netw. **37**, 256–282 (2016). https://doi.org/10.1016/j.adhoc.2015.08.023

119. B. Liu, H. Chen, Z. Zhong, H.V. Poor, Asymmetrical round trip based synchronization-free localization in large-scale underwater sensor networks. IEEE Trans. Wirel. Commun. **9**(11), 3532–3542 (2010). https://doi.org/10.1109/TWC.2010.090210.100146

120. D. Pompili, I.F. Akyildiz, A multimedia cross-layer protocol for underwater acoustic sensor networks. IEEE Trans. Wirel. Commun. **9**(9), 2924–2933 (2010). https://doi.org/10.1109/TWC.2010.062910.100137

121. R. Ashri, H. Shaban, M.A. El-Nasr, A novel fractional fourier transform-based ASK-OFDM system for underwater acoustic communications. Appl. Sci. **7**(12) (2017). https://doi.org/10.3390/app7121286

122. R. M. Ashri, H. A. Shaban, M. A. El-Nasr. *BER of FRFT-Based OFDM System for Underwater Wireless Communication*. In 33rd National Radio Science Conference, NRSC 2016 (2016), pp. 266–273. doi:https://doi.org/10.1109/NRSC.2016.7450837

123. M. Jouhari, K. Ibrahimi, M. Benattou, *Implementation of bit error rate model of 16-QAM in aqua-sim simulator for underwater sensor networks* (Springer, Berlin, 2017). https://doi.org/10.1007/978-981-10-1627-1_10

124. H. Alhumyani, R. Ammar, R. Elfouly, A. Alharbi. *Heurstic Approaches for Underwater Sensing and Processing Deployment*. 11th International Computer Engineering Conference (ICENCO 2015) pp. 86–91 (2016). https://doi.org/10.1109/ICENCO.2015.7416330

125. M. Jouhari, K. Ibrahimi, H. Tembine, M. Benattou, J. Ben Othman, Signaling game approach to improve the MAC protocol in the underwater wireless sensor networks. Int. J. Commun. Syst. (2019). https://doi.org/10.1002/dac.3971

126. N. Krishnaraj, M. Elhoseny, M. Thenmozhi, M.M. Selim, K. Shankar, Deep learning model for real-time image compression in Internet of underwater things (IoUT). J. Real-Time Image Proc. (2019). https://doi.org/10.1007/s11554-019-00879-6

127. H. Bennouri, A. Yazidi, A. Berqia. *A Pursuit Learning Solution to Underwater Communications with Limited Mobility Agents*. In Conference Research in Adaptive and Convergent Systems, RACS 2018 (2018), pp. 112–117. doi:https://doi.org/10.1145/3264746.3264798

128. M. Ammar, K. Ibrahimi, M. Jouhari, J. Ben-Othman (2018). *MAC Protocol-Based Depth Adjustment and Splitting Mechanism for UnderWater Sensor Network (UWSN)*. In IEEE Global Communications Conference Proceedings, (GLOBECOM 2018). https://doi.org/10.1109/GLOCOM.2018.8647644

129. E.M. Khater, D.M. Ibrahim, M.T.F. Saidahmed. *Contention-Based MAC Protocol in UWSNs: Slotted-CS-ALOHA Proposed Protocol*. 11th International Conference on Computer Engineering and Systems, ICCES 2016 (2017), pp. 73–78. https://doi.org/10.1109/ICCES.2016.7821978

Chapter 10
Health Informatics as a Service

P. M. Rekha and M. Dakshayini

10.1 Introduction

The healthcare industry is sensitive and delicate, which makes it a field that requires precision and objectivity. Various researchers have surveyed in the healthcare setting, which supports the adoption and prevalence of healthcare informatics. Cognitively, the research done by the researchers brings forth theories such as the Big Data Theory and Orem's Theory. The approach seeks to address what the healthcare should do for its efficacy with respect to the clinicians who include nurses. Notably, information technology (IT) personnel who are proficient in electronic media records (EMR) use coded language which has various acronyms that cannot be easily decoded by a non-IT individual such as a physician, doctor, or clinician. Arguably, the coherent working of healthcare professionals and developers in healthcare informatics is crucial. Healthcare professionals should help in the building of systems supporting EMR, for example, to aid in building a comprehensive and custom health informatics system for the institution they work for. Integration of cloud technologies for computer management systems has become a challenge in the health sector and attracts contentious debates across the globe. The healthcare industry should care about this challenge since the world is changing, populations are increasing, and the need for quick and excellent

P. M. Rekha (✉)
Department of Information Science & Engineering, JSS Academy of Technical Education,
Bangalore, Karnataka, India
e-mail: rekhapm@jssateb.ac.in

M. Dakshayini
Department of Information Science & Engineering, BMSCE, Bangalore, Karnataka, India
e-mail: dakshayini.ise@bmsce.ac.in

© Springer Nature Switzerland AG 2020
F. Al-Turjman (ed.), *Trends in Cloud-based IoT*, EAI/Springer Innovations
in Communication and Computing, https://doi.org/10.1007/978-3-030-40037-8_10

performance is needed [1]. Failure to adjust to changing society and population needs will have some health institutions thrown out of their practice. It is important to note that healthcare institutions are primarily businesses just like other enterprises which harbor professionals as their employees. To that accord, the healthcare industry should exercise due diligence and strive to integrate cloud technology for efficient health informatics effectively. This will ensure clients remain loyal due to outstanding service delivery which will see health institutions build their reputation and reach a broader client list.

10.1.1 The Relevance of Cloud Computing

Examples of cloud computing technology management systems, which are also a proponent of nursing informatics, include electronic health records, electronic assisted services, and digital leadership. Further, a case study of a healthcare organization known as MEDITECH shows that it seeks to grow through healthcare informatics' investments. MEDITECH needs an expansion of its systems, aimed at developing more space for electronic medical record (EMR) while improving the efficiency of patient-handling. The company has adopted cloud computing to improve its software design meant for the healthcare industry. The move is informed by a fast-changing technology world which necessitates upgrades in the healthcare industry. The company reiterates that healthcare institutions that do not adopt cloud technologies are missing facilities and in due time they might become obsolete. Improved and rejuvenated medical systems that accommodate cloud technologies are beneficial as they cater better for patients whether inpatient or outpatient and sometimes home-patient on-demand services in healthcare organization.

10.1.2 Hypotheses of the Relationship Between Health Informatics and Cloud Computing

Comprehension of the technical terms used by nurses and technical developers is also a problem that is a gap in healthcare informatics. Collective and inclusive training of healthcare professionals as well as developers will be of the essence as it will iron out the technicalities and ambiguity of technical and healthcare jargon. Cases of the loss of patient data have been prevalent which raises the question of viability and reliability of the health informatics systems. Failure of collaborative working between developers at MEDITECH and healthcare professionals has been identified as the main cause of data loss in healthcare institutions that have purchased its electronic systems. It emerged that there were predominant challenges when nurses were handling the electronics systems because the developers failed to address the former needs such as through orientation and training. The prevalence of technical jargon resulted in the poor decoding of orders. Importantly, the costs incurred while adopting cloud technology are a concern for matters of long-term benefits. The cloud technology benefits outweigh the cost incurred in its adoption in healthcare.

10.2   Literature Review

10.2.1   The State of Cloud Computing in Healthcare

According to Alexander and Fields, 2014 integration of cloud computing for computer management systems accounts for 30% of the problems facing the healthcare industry. This is the most significant percentage compared to other individual difficulties of healthcare informatics which include inadequate financial resources at 26%. Alexander and Fields continue to assert further that a healthcare setting in the 21st century that has no electronic incorporation or limits it is doomed to fail. The reason for failure is that different conditions are also mutating with the changing environment. Research has gone further to incorporate computer systems and so should the healthcare industry such as in nursing do to apply knowledge that the researchers have developed [2]. Stakeholders affected by this issue include healthcare institution owners, clinicians, patients, nursing bodies, and medical facility suppliers. These people expect the nursing profession to reward them handsomely [3]. That is not possible if the maximal potential is not reached and one of the ways of reaching maximum potential is through integrating computer management systems which are healthcare informatics through the inclusion of cloud computing. Responses to patients in emergency situations become swifter than in the analog or manual maxim, which is called on-demand cloud computing. Treatment of patients becomes faster and accurate; this sees the stakeholders rake in more profits, expand their institutions, build their reputation, and advance their careers. Failure to fix the issue, all of the above will not be achieved.

10.2.2   Motivators of Cloud Computing in Healthcare

Digital leadership is affected by the flexibility to adapt to new environments and the dynamism the information technology world attracts. Leadership is subject to health informatics as it determines how the healthcare setting would be run, be it an ambulatory health clinic or a normal patient ward. Some leaders remain conservative and affect the necessary change management needed to improve their healthcare unit. Notably, such kind of leaders frustrates moves from their juniors or subordinates to effect change [4]. A classic example is when a leader is out for his or her errands, and an emergency occurs, and his or her expertise or authority is required, but there is no apt mode of communication. Cloud computing comes in handy for telehealth and virtual care through technologies such as video conferencing, which can help address such a situation.

Orem's theory suggests that self-care is crucial in healthcare as it aims to fix inherent deficits. This approach comes in convenient to address how a healthcare situation can be solved. The theory of self-care is self-definitive, it requires the person in charge of a healthcare situation to be entirely in charge and competent in ensuring lives are saved, or rather efficacy is achieved when handling patients and resources at a medical institution [5].

Strategically, the kind of leader seen above should be able to embrace technology such as that of video conferencing that allows hospital rooms to have monitors that can be used for ease of communication through cloud computing [6]. This will also enable the leader to see the kind of situation at the medical institution which will intuitively help him or her address it. Self-care deficit theory requires a professional to be responsible for his or her health as well as that of others. As such, the leader should critically adopt the proposed technologies as that will save lives. Moreover, the theory of nursing systems insists that a clinician can meet patients' needs through an intelligent application of knowledge which health informatics provides. Cloud technology in healthcare can be made for public use by building data center for the hospitals which become less costly especially for service provision agility. People and hospital network can access the different medical data on the cloud with ease due to the convenience of cloud computing. Furthermore, cloud technology in healthcare can be used for private clinical applications, and non-clinical applications such as management of revenue cycle management. Cloud technology also eases patient management task such as billing and claims. Servers, networks, and storage are synchronized by cloud technology to enhance automation and seamless access for the healthcare industry.

10.2.3   The Efficiency and Applicability of Cloud Computing in the Handling of Health Care Tasks

Electronic health records (EHR) are crucial in the current world. Cloud platforms can be hosted internally or externally; the former is more secure. They serve different purposes which see them improve efficiency and effectiveness in healthcare units. A unit that embraces electronic health records on a cloud platform saves itself from the agony of losing lives as well as confusion especially in emergency situations due to easy and on-demand access of medical and patient information [7]. Some clinicians see that adopting electronic health records is technical while others fear to lose their jobs. EHR's offer healthcare more rewards than harm, and as such clinicians should not feel worried by the same. EHR's help clinical institutions develop quick responses and attention to patients [8]. Other gains brought by cloud computing include automated refill ordering monitoring supply chain in real time, event-based alerts and data logging and AI decision-making. Transparency and traceability are enhanced in the healthcare supply chain which makes operations more flawless and seamless. Event-based alerts and data logging ensure that everything done on any machine connected to the cloud leaves footprints that can be used for audit. AI decision-making is a proponent of business intelligence which will help healthcare professionals make informed decisions through thorough and advanced analysis of medical data. Things that can be analyzed through the efficacy of AI on cloud include resources and patient ratios, operational gaps, and business development opportunities.

It is retrogressive to note that some health institutions do not allow for the integration of their records with other health institutions [9]. Maybe, the law does not explicitly require every health institution to cooperate with each other, but this context thinks of it as a necessity. Some of these institutions assert that other companies might hack into their systems and retrieve their secret information which according to them is unethical [10]. However, as much as there are risks of losing information healthcare institutions should strengthen their electronic security and cooperate seamlessly with other organizations through cloud computing. Social network research provides medical information on a cloud platform which makes it easy to access data for research purposes. HIPAA ensures that healthcare use patient information responsibly and is further protected by general data protection guidelines.

Electronic assisted services also define how a healthcare institution integrates computer management systems which allow for uniformity of medical records through cloud technology. Institutions that have engaged electronic medical facilities can enter data from any branch and have the details recorded in real time allowing for no double entries or mechanical errors which makes information available uniform. Cases of a hospital branch having different records from that of another concerning the same patient or patients will be eliminated. It is vital that healthcare institutions entirely transform into digital masterpieces through cloud technology as this will steer them to excellence and credibility [11]. Today, some machines can scan a person's body and record as many ailments that a person might have by sending information to, and retrieving data from a cloud platform; this is one of the much progress that integration of computer systems has brought to the healthcare setting [12]. It helps to remove the uncertainty that is carried by the manual detection of diseases with the typical look at symptoms. Primarily, some health institutions stick to old machines which over time become ineffective, the authorities in the same institution become reluctant and do not release funds for upgrade or maintenance that can allow for cloud technology adoption or advancement in their institutions [13].

10.3   Cloud Computing Application in Healthcare

The architecture of cloud computing application in Healthcare system with interconnected different components are depicted in Fig. 10.1.

10.3.1   Data Security

Service Provider

With reference to service provision for healthcare applications, cloud computing
services are offered, mostly, by corporations with experience in data management
and security. The use of health informatics as a service allows healthcare institutions
to concentrate on creating reliable service for their users while letting an independent entity handle the technical part of information access. Information access
for institutions and their users is controlled by a policy that the institutions have
complied and implemented with respect to the user-provider agreement. In this

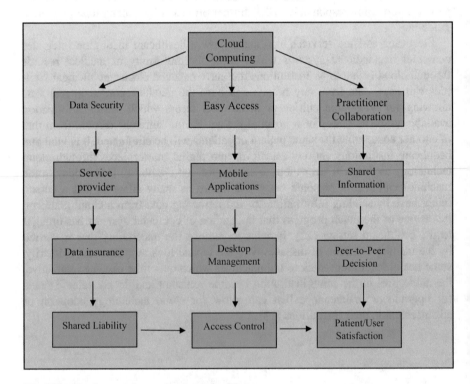

Fig. 10.1 Cloud computing application in healthcare

case, the selection of a suitable cloud computing service provider depends on the understanding of the policy definition with respect to data confidence and security.

Data Insurance

Health informatics is specific in the type of data and information accessible at any given moment. For users, health informatics provides a historical analysis of treatments and interventions and also Metadata on the associated risks of healthcare access. For the practitioners, only information shared between a user and the institution, in general, can be accessed with appropriate consent from the user. Breach to the information database can be detrimental to legal consequences for both the service provider and the corporations/institutions using the service. In response to this approach, the experience of the service provider and the mechanisms put in place to manage data and to protect clients serve as data insurance. The need for healthcare institutions to hire cloud computing services aims at achieving shared liability.

Shared Liability

The cost of a data breach can span out of the budget limits of emergency expectations for most healthcare institutions. Healthcare institutions are mainly targeted for their hoarding of user information. Users of healthcare services provide a wide range of personal information including healthcare insurance information. This data richness influences hackers and other unauthorized users to seek information that can benefit them in two significant ways. Firstly, sensitive personal information is traded for financial gains and access to this data is in high demand. Secondly, users can be blackmailed for specific favors using personal data held in health institutions' databases. With the responsibility for protection handed over to service providers, issues to do with data security are the payable services as the providers have considerable experience in retaining their customers by providing up-to-date security protocols on data management.

10.3.2 Ease of Access

Mobile Applications

For users and practitioners, ease of information access is among the benefits cloud computing offers to healthcare sector in general. Particularly, practitioner-patient association and collaboration to achieve the best intervention outcomes are among the benefits of using mobile applications in providing medical care. Cloud computing provides real-time association between practitioners and their patients through mobile applications that allow information access and updating on a remote basis.

Desktop Management

Most cloud computing services function through Web 2.0 where data collaboration and sharing has shared benefit. Healthcare institutions require functional and reliable service that offers the flexibility of upgrades and changes in the institutional web eco-system. The organization of data differs from one institution to another and also required different programming and computing strategies to achieve the needed goals. In this regard, cloud computing allows desktop and remote management of databases through collaborative association of the service provider and the health-care institution. Management changes within healthcare institutions are influenced by policy changes effected by authorities or those internally generated.

Access Control

Access control is mostly a feature of cloud computing that correlates to patient-practitioner ethics of association. With cloud computing, a user has the authority to provide consent to the type of information needed for access by any given practitioner. Thus, for practitioners interested particularly on checkups for health concerns do not need to provide full access to their medical health records. By offering this option, unlike with traditional systems, users of healthcare feel safer controlling the access of their data that they would be giving the institutions the full control of their data.

10.3.3   Practitioner Collaboration

Shared Information

Cloud computing is not only beneficial for practitioner-patient tasks but also for peer-to-peer collaboration. Healthcare databases are created to offer practitioners with referencing sources that relate to current and historical health interventions. They also provide clinical breakthroughs that are applicable to various emerging care settings. In this regard, the sharing of information among practitioners aims at providing standard and ethical interventions that correlate to specific medical conditions. Mostly, new and experienced care providers require reference materials for their responses to be meaningful for various clinical situations and to achieve the desired outcomes, access allows for prompt, ethical, and standard interventions.

Peer-to-Peer Decision-Making

The differences in expertise require collaborative decision-making in clinical settings. With cloud computing, shared information between practitioners helps on the management of medical outcomes. A variety of health outcomes are associated with the interventions provided and the interaction of medicinal alternatives. Without collaboration on these decisions, a cardiologist without the proper information about the outcomes of chemotherapy session may not make the right decision, and in the process, the unfavorable interaction of both interventions may result to medical complications to the patient. Thus, peer-to-peer decision-making is an essential practitioners' strategy powered by prompt cloud computing communication and data sharing.

Patient and User Satisfaction

The overall application of cloud computing in healthcare is to achieve reliable and quality care for users of the system. Healthcare institutions aim at achieving efficiency and to retain users of the system. By integrating data security, ease of access, and practitioner collaboration, health informatics powered by Web 2.0 service provision benefit considerably from cloud computing. The need to achieve patient/user satisfaction by offering quality service can only be achieved by combining data security, standard decision-making applications, and ease of information access.

10.4   Methodology

10.4.1   Instrument Development

Secondary sources of data were used to inform the work done here. Journal articles, books, and medical websites will facilitate the provision of detailed information that will guide the findings, discussion, and conclusion of this research [14]. The sources are not older than 6 years to ensure relevance and applicability of the information received from them. Healthcare industry trends will be observed to ensure relevance and applicability of the developed knowledge through this research. Expansion of healthcare informatics will be based on prioritization of needs as well as the magnitude of identified. The essence of prioritization will be the smooth implementation and completion of laid plans.

10.4.2   Data Collection Procedure

Interviews, observation, reading, and recording of information were the procedures used for this research. Ethical considerations such as respect for anonymity were taken care of for interviewees who acted as primary resources. Open-ended questions were given to ensure the answers given were detailed and as possible. Information collected from secondary sources are tabulated.

10.4.3   Sample Demography

Interviewees were selected through random sampling. In this work, nine respondents from three different cities were selected. Each city had three respondents. The respondents were chosen from three age groups 23–30, 31–44, and 45 and above. There was no division of gender in the interviews as that would be discriminative; answers from either male or female respondents were treated as representative of their age group. Information received from the different sources were tabulated for a more visual and dominant appeal.

10.5   Data Analysis and Discussions

10.5.1   Measurement Model Analysis

The answers provided by the interviewees were weighed against each other in terms of proponents and opponents. The information derived from secondary sources are presented in Table 10.1 which represents challenges and solutions.

10.5.2   Structural Model Analysis

Adoption levels were also documented and are presented as shown in Fig. 10.2. The graph shows the adaptation levels of healthcare informatics in cloud computing over time. Adoption rate may increase in the range of 45–50% with respect to infrastructure, leadership, and security by 2020.

Interviewee's answers were also tabulated for a clearer perspective and comparison. The structural model analysis noted disparities in responses due to age differences as well as physical locations. For instance, individuals from San Francisco were more adamant in the adoption of cloud computing in healthcare compared to individuals from Los Angeles and New York. One can tell that the individuals from San Francisco understand cloud computing more than their

Table 10.1 Summary of secondary results and findings

N = Age	Challenges	Solutions
11.	Low adoption levels of cloud technologies for health informatics	– More funding – Awareness and training – Research and development
22.	Slow lags in patient care	– Adoption of health informatics systems by providing easy access to information through the cloud
33.	Gaps in digital leadership	– Provision of healthcare facilities that can allow for remote access and consultancy; telehealth and virtual care made possible by cloud computing
44.	Security fears	– Strengthening of health informatics systems by engaging security professionals, for instance by enabling factor authentication to increase access control and security on cloud platforms
55.	Aging infrastructure	– Need to purchase modern facilities to avoid obsolescence in the dynamic and competitive health informatics world

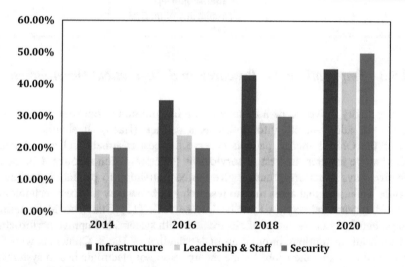

Fig. 10.2 Healthcare informatics adoption

counterparts, more so, it is part of their culture as tech-oriented persons. The young respondents saw the need for adopting cloud technology because they are well versed with the developments in the technology world. The age group 31–44 are also knowledgeable about cloud technology; they are people in the productive age and understand the real benefits of cloud technology in healthcare. The last age group of 45 and above shows less conviction for cloud technologies in healthcare; this can be attributed to their longtime operating under analog healthcare settings, and they were convenient for them. The individuals with higher age group as depicted in Table 10.2 are highly conservative and showed less support for cloud technology in healthcare.

Table 10.2 Interviewee responses

Age	City		
	New York	Los Angeles	San-Francisco
23–30	Cloud computing would be an added advantage in healthcare	The world is increasing cloud technology so need to utilize cloud technology	Immense need to adopt cloud technology in healthcare
31–44	Need to cope up with the world in matters technology	You cannot ignore technology; it brings more harm than good; hence there is need to adopt cloud technology in healthcare	Immense need to utilize cloud technology in healthcare
45 and above	Adoption of cloud technology should be done but not as a matter of urgency	Adoption of cloud technology can wait, there are other issues in healthcare such as affordability and availability of medical care	Immense need to adopt cloud technology in healthcare

10.5.3 Implications for Research and Theoretical Development

Orem's theory proves to be a remedy to the integration of computer management system for adroit health informatics as a service (HaaS). It should be noted that as per Orem's theory, patients deserve the best treatment and the clinicians should strive to offer the best of services as their profession requires. One of the effective ways of ensuring nurses give the best attention to patients is by having flexible leadership that takes time to research and know how different technologies will aid service delivery in their healthcare units [15]. Staff should be trained comprehensively on the usage of electronic health systems to improve performance in healthcare institutions through an understanding of HaaS. Clinicians who feel threatened of losing their jobs by the incorporation of electronic health systems in cloud computing should be educated on the benefits of EHR. Staff members who do not meet the criteria set for personnel responsible for handling electronic health records should be reshuffled and assigned roles that connote their competence [16].

Implementation of the above strategy can be achieved through the evaluation of the strengths of different clinicians; each clinician is moved to his or her area of expertise. Funding is needed to facilitate training for clinicians who show interest in the integration of computer system management on cloud platforms. Rewards should be set to motivate staff into the new culture. This will make it a worthy course. Importantly, measures of performances should be introduced to ensure the implementation process happens as planned [17]. Benchmarking would serve as a proper measure of performance that would see the apt completion of the implementation process. One ethical aspect related to an evaluation of people's strengths to determine their place regarding the integration of computer systems is employee retention.

Employees who are clinicians in this context might lose their jobs if the new requirements do not match their abilities or strengths [18]. From an ethical point of view, loss of employment will result to deterioration of living standards which can lead to criminal or immoral activities. As such, this strategy will strive to considerably reshuffle the clinicians who show strengths in other departments to the same departments [19]. In the future, there is a dire need to research how the extremely dynamic technological world will ensure the vocations of clinicians do not become redundant due to the dominance of advanced medical machinery. This is because the traditional clinician has a way that he or she is used to; a change that necessitates acculturation might be unacceptable to clinicians. Training might be provided, but passion and conventions define an individual; therefore when some clinicians are not malleable to the dynamics of the technocratic world, then research is of the essence to help.

Most respondents showed a great need for the adoption of cloud technology; it was clear that they saw more benefits than harm from its inclusion in healthcare. Cognitively, the Big Data theory is about the aspect of getting massive amounts of information and then synchronizing into analytical bits that form real-time records that can be used for patient care at any medical facility with supporting cloud technology systems [20]. That makes it easy to see the medical records of a patient upon reaching a medical institution. That can also help in emergency cases where a real-time and fast check can be done to know the medical record of a patient and how best to attend to them as seen in the findings section.

10.5.4   Limitations and Further Research

There is a new piece of knowledge of nursing informatics that has been generated which is called mHealth and uHealth [21]. That enables prompt self-care and personal monitoring activities through the cloud. The findings relate to the existing big data theory and knowledge of health informatics in a professional and interactive scope, but there are challenges. Health informatics has seen the development of systems that can initiate the generation of new knowledge from old knowledge, which is progressive. Technology has been embraced as it continues to be dynamic. That can be seen in the big data theory that applies to the case here. The large pool of data on cloud platforms has enhanced on-demand data access and real-time patient care attendance and delivery. The challenge in the big data as pertains big data theory is that there are flaws in doing the remote real-time simulation due to lack of cloud technologies that would have eased a coordinated and centralized posting of data from different branches. Therefore, more research needs to be done to develop workable real-time remote simulations for cloud efficiency for each healthcare institution's custom need and operational model.

10.5.5 Implications for Practice

Cloud computing for electronic resources such as monitors which are computer-oriented should be put in place across medical institutions to help improve patient and healthcare staff experience. All this will be crowned by the strict implementation of the strategic move of placing clinicians in their departments of strength; this will ensure the integration of cloud computing electronic systems becomes successful. Importantly, adoption of cloud services such as Oracle 12c, Amazon Web Service, or Azure will give healthcare institutions more leverage against their competitors as well as increased capacity of securely handling large data [22]. Electronic health records (EHRs) as data source is more advantage if patient's information are combined, aggregated, and analyzed to ease the group of valuable data [23]. The authors-initiated issues related to distressing hospital Evidence Base Management and recognized six evidence causes that healthcare administrators can practice in evidence-based decision-making processes [24]. Deep learning techniques in medical signals and images can help the clinicians for decision-making. Deep Learning applications in healthcare protect a wide range of problems ranging from cancer screening and monitoring of disease with suggestions for custom-made treatment suggestions [25].

Data analytics will further make health informatics a seamless discourse where clinicians can predict patient behavior and outcomes, making the healthcare industry more versatile and efficient. Lastly, healthcare professionals will become more skilled and proficient in their delivery of service through training and daily application of healthcare informatics through on-demand and reliable cloud access. The downside of widespread adoption of cloud computing in healthcare informatics is unemployment; unskilled personnel might find their roles redundant. To avoid such misfortunes, healthcare professionals will have to adapt to the change through competence and skill training to ensure they are always ahead of change.

10.6 Conclusion

Inferentially, the healthcare industry requires adroit measures that effect change to guarantee the healthcare industry the success it strives to achieve. The world is changing; technology is taking over and need for improved performance is required by the populations and healthcare institutions alike. Patients need swift but sure ways of their treatment, they need attention and care that is exceptional, and that assures them of their good health. Healthcare institutions have not fully embraced cloud technology in healthcare informatics, but there is room for growth. Stakeholders, on the other hand, want their investments or interests in the medical field assured through excellence in healthcare institutions. Therefore, integration of cloud computing in their computer management systems will become a cornerstone and pillar of excellence through the adept incorporation of Orem's theory that advocates for the improvement of efficiency. Electronic health records on cloud platforms are essential to the healthcare setting and require clinicians to be proficient in their use. This should be met with delight by every healthcare institution because the patient is the primary reason why the institutions exist and the healthcare industry thrives. Cloud computing in healthcare informatics is the future for a real revolution of the healthcare industry.

A.1   Appendix: Definition of Technical Terms

MEDITECH: A software vendor that offers healthcare informatics applications.
Azure: Microsoft's cloud technology.
Amazon Web Service (AWS): Amazon's cloud technology.
Oracle 12c: Oracle's cloud technology.

References

1. B. Greenwood, *Problems with Nursing Informatics* (Olympia, London, 2013)
2. D. Alexander, W. Fields, *Challenges within Nursing Informatics and Health IT* (McGraw-Hill, Chicago, 2014)
3. V.K. Nigam, S. Bhatia, Impact of cloud computing on health care. Int. Res. J. Eng. Technol. **3**(5), 2804–2810 (2016)
4. S.G. Taylor, *Orem's Self Care Deficit Nursing Theory* (Sage, Washington, 2014)
5. S. Misra, A. Samanta, Traffic-aware efficient mapping of wireless body area networks to health cloud service providers in critical emergency situations. IEEE Trans. Mob. Comput., 1–1 (2018)
6. C.P. Kumar, *Application of Orem's Theory* (Pearson Education, San Francisco, 2012)
7. R. Ratwani, T. Fairbanks, E. Savage, K. Adams, M. Wittie, A. Boone Gettinger, Mind the gap. Appl. Clin. Inform. **7**(4), 1069–1087 (2016)
8. F.C. Manzini, J.P. Simonetti, *Nursing Consultation Application to Hypertensive Clients* (The American Psychological Association, New York, 2012)
9. A. Conte, R. Quattrin, E. Filiputti, R. Cocconi, A. Luca, et al. Promotion of flu vaccination among healthcare workers in an Italian Academic Hospital: An experience with tailored web tools. Human Vaccines & Immunotherapeutic **12** (2016). https://doi.org/10.1080/21645515.2016.1186319
10. F.A. Sampaio, *Application of Theory to Nusring Practice* (Xulon, Minneapolis, 2013)
11. T. Berry. (2012). http://www.mplans.com/articles/how-to-perform-a-swot-analysis
12. Weberience LLC. (2014). http://pestleanalysis.com/swot-analysis-in-healthcare/
13. W. Gretzky. Strategic planning and swot analysis. In J. P. Harrison, *Essentials of Strategic Planning in Healthcare* (2013). pp. 92–96
14. K. Pinaire, S. Sarnikar, A quantitative approach to identify synergistic IT portfolios, in *Reshaping Society through Analytics, Collaboration, and Decision Support. Annals of Information Systems*, ed. by L. Iyer, D. Power, (Springer, Berlin, 2015), pp. 135–156
15. S. Toromanovic, Nursing information systems. J. Inform. Sci., 35–47 (2013)
16. Y. Al-Turjman, K. Ever, E. Ever, H. Nguyen, D. Deebak, Seamless key agreement framework for Mobile-sink in IoT based cloud-centric secure public safety networks. IEEE Access **5**(1), 24617–24631 (2017)

17. N. Wickremesinghe, R. Arias, J. Wilgus, C. Gonzalez. Reducing healthcare disparities with Technology. In *Encyclopedia of Information Science and Technology* (2015), pp. 3419–3427
18. C. Houston, *The Impact of Emerging Technology on Nursing* (Greenleaf Group. LLC, Austin, 2013)
19. F. Al-Turjman, M.Z. Hasan, H. Al-Rizzo, Task scheduling in cloud-based survivability applications using swarm optimization in IoT, Trans. Emerg. Telecomm (2019). https://doi.org/10.1002/ett.3539
20. B.T. Basavanthappa, *Nursing Theories* (Dorrance, Philadelphia, 2013)
21. S.P. Firminger, J. Garms, R.A. Hyde, E.K. Jung, C.D. Karkanias, E.C. Leuthardt, J.D. Inaldo Jr., *U.S. Patent No. 9,892,435* (U.S. Patent and Trademark Office, Washington, DC, 2018)
22. O. Blakstad. (2013). https://explorable.com/research-designs
23. S. Marceglia, P. D'Antrassi, M. Prenassi, L. Rossi, S. Barbieri *Point of Care Research: Integrating Patient-Generated Data Into Electronic Health Records for Clinical Trials*, AMIA Annual Symposium Proceedings (2017), pp. 1262–1271
24. A. Janat, E. Hasanpoor, S. Hajebrahimi, H. Sadeghi-Bazargani, Evidence-based management-healthcare manager viewpoints. Int. J. Health Care Qual. Assur. 31(5), 436–448 (2018)
25. F. Al-Turjman, H. Zahmatkesh, L. Mostarda (2019) *Quantifying Uncertainty in Internet of Medical Things and Big-Data Services Using Intelligence and Deep Learning*. IEEE Access

Chapter 11
Cloud-Based Smart IoT Architecture and Various Application Domains

Debashis Das, Sourav Banerjee, and Utpal Biswas

11.1 Introduction

The Internet of Things (IoT) [1] can always be regarded as just a sort of ecosystem in which all physical entities, populations, and creatures have such a distinct identity and can pass information across the web without communication. IoT has become a mix of various techniques. It developed the Internet, wireless, and micro-electromechanical systems (MEMS). These terms can also be regarded as the Internet of Everything. The premise of IoT is that things and objects communicate and swap huge-scale data. Nowadays, various industries use IoT equipment to gather real-time and ongoing information and create stronger company choices to boost customer satisfaction. A company needs to collect information produced from the IoT and this information is growing exponentially; it forces to give attention to cloud computing to manage IoT information. The cloud becomes a significant option to store IoT information, as it is expensive and sensitive to multiple organizations to keep this information in the cloud. It has benefits than on-premises space to keep IoT information. First, there is a definite link between appliances and the supplier of the public cloud. Therefore, this direct connection allows information to be stored more quickly; thus it requires small space and reduced price per device. Second, the cloud provider problem is data management and storage management, so the organization only has to use the service. The paradigm of the IoT is focused on smart, self-configuring things interlinked in a complex, worldwide network environment [2]. IoT is the most emerging technologies that render the computing environment

D. Das · U. Biswas
University of Kalyani, Kalyani, India

S. Banerjee (✉)
Kalyani Government Engineering College, Kalyani, India
e-mail: mr.sourav.banerjee@ieee.org

© Springer Nature Switzerland AG 2020   199
F. Al-Turjman (ed.), *Trends in Cloud-based IoT*, EAI/Springer Innovations
in Communication and Computing, https://doi.org/10.1007/978-3-030-40037-8_11

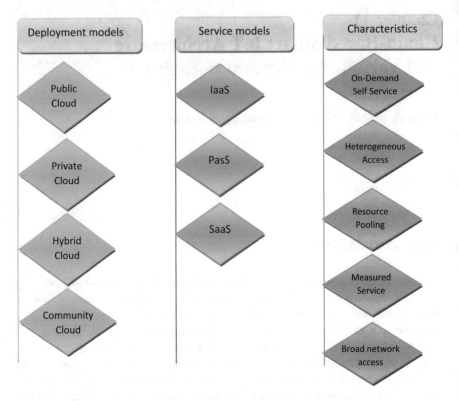

Fig. 11.1 Cloud computing deployment models, service models, and characteristics

pervasive and ubiquitous. IoT generally relates to the actual environment and restricted memory and execution capacity of small objects, as well as the significant issues of performance, flexibility, safety, and security.

There are various proposed definitions for describing cloud computing. National Institute of Standards and Technology (NIST) has given a simple definition; cloud computing is a framework to access data of shared pool for enabling on-demand, ubiquitous service of computing resources as like storage, servers, networks, etc., which can be managed and processed with minimum effort [3]. Figure 11.1 has shown four types of deployment models, three service models, and five key characteristics in the domain of cloud computing.

In the public cloud [4] environment information is available over the Internet. The users of a specific organization can access information in the private cloud domain. The information in the private cloud [4] is secured than the public cloud. Hybrid model is the combination of the private and public cloud models. As there are different possible drawbacks for consumers, by choosing the hybrid model [4] it can be made possible to overcome these limitations. The community cloud [4] is a special type of cloud model where a group of users forms a number of organizations that can interact and share information with them. Cloud

Table 11.1 Comparison of cloud computing with the Internet of things [43]

Properties	Cloud computing	Internet of things
Nature	It is ubiquitous	It is pervasive
Resources	Resources are virtual	Resources are real
Storages	Unlimited	Limited or not available
Integration	Uses the Internet for services delivery	Uses inter for point of consolidation
Big data	It is for managing big data	It is a source of big data
Processing abilities	Unlimited computational capabilities	Limited computational capabilities
Accessibility	Resources are available from everywhere	Things are available everywhere at any time

computing provides efficient, on-demand, and highly available networks access to various configurable computing assets. Cloud computing has innumerable storage capabilities and processing power, which is one of the most mature technologies that can at least solve IoT problems by extending it. IoT and cloud computing integration is known as a cloud of things [5]. This is a novel initiative technology where Cloud Based Internet of Things (CBIT) is being combined with the cloud and IoT. CBIT solves some problems such as IoT limitations, analysis of data, and computing and creates new opportunities like Things as a service [6] and Smart Things. However, the cloud and IoT have their own complementary properties, which are shown in Table 11.1. Combing both these technologies we can give a new Internet world for the future in order to change today's Internet direction.

In recent years, unmanned aerial vehicle (UAV, known as drone) applications are increasing day by day with significant attention. This affordable and available technology is constantly being improved and rendered to a range of novel applications worldwide. Originally considered to be used for their military purposes, UAVs are now used by private contractors, small businesses, and large companies for various additional tasks. UAVs are a part of the cloud when integrating them with the cloud network. As a consequence, the cloud network includes the efficient cloud-based servers for data and computations. On the other side, UAVs provide services and resources for the physical environment. Cloud computing expands for interactive UAVs with many possibilities. UAV monitoring and application usage is feasible from everywhere due to the pervasive value of cloud computing.

11.2 Literature Review

IoT has emerged through the success of accessibility and integration, which embraces ineffable computations, where everything will be linked. Two recent techniques, IoT and cloud computing, are expected to increase their huge use and acceptance. IoT has become so widespread where it is becoming a problem to deal with the information generated by its components. One alternative to this issue is cloud computing. However, it doesn't seem to have many advantages to the concept of combining IoT and cloud. The factors for CBITs creation and the main inclusion problems are also discussed. Stergiou et al. [7] have presented a survey on IoT and cloud computing with a concentrate on both technology safety problems. In particular, they merge the above two techniques (i.e., cloud computing and IoT) to explore popular characteristics and to find out the advantages of their inclusion. Ray [8] observed on IoT web applications in the context of the various business fields such as software design, device management, system managing, heterogeneity governance, data management, evaluation instruments, implementation, and tracking. Zahu et al. [9] have presented an architectural design and specific security and data protection specifications for future mobile CBIT systems, identifying ludicrousness from most of the previous work, and addressing demanding concerns of efficient packet transfer and flexible authentication protection by suggesting the new advanced confidentiality protection of information without a holomorphic public key. Diaz et al. [10] have provided an overview of regions of implementation: cloud applications, cloud environment, and IoT middleware. In fact, some suggestions for involvement and analytical information are examined, as well as various difficulties and accessible analysis questions are highlighted. Sivakumar et al. [11] analyzed in detail CBIT applications to contrast cloud and IoT applications with architecture designs, protocols, system designs, database technologies, detectors, and model improvisation algorithms. Belgaum et al. [12] described the fields, advantages and difficulties of CBIT applications that are accessible to study in those creative frameworks are demonstrated. Auger et al. [13] have experienced CBIT Platform architecture and growth. Such schemes present fresh problems for studies, in specific for the compilation, treatment, and usage of observations. This paper [14] highlights the need for cloud and Internet integration of things, an agent-based cloud that uses a cloud IoT paradigm based on the layered citation design. In [15] they have analyzed and debated how these problems have to be integrated and how they are addressed in the literature. In [16] the author discussed accessible difficulties and feasible alternatives for the Internet for the future. Chowdhury et al. [17] evaluated and differentiated the to present by using proper methods to ensure essential safety requirements in IoT communications. The CBIT architecture was suggested by Silva et al. [18] to enhance the implementation of intelligent, remote- and control-based manufacturing technologies. They have also suggested the architecture of an integrated inductive engine situation by using particular techniques accessible for a utility. Elazhary et al. [19] have presented IoT-related technology such as all-embracing and computational technology, the Internet of Nano Things (IoNT) [20], and the Internet of Underwater Things (IoUT) [21]. The main focus of this article is the inclusion of the cloud and IoT paradigms and their use situations. However, there is no comprehensive study of the current cloud IoT paradigm, involving entirely fresh features, advantages, difficulties, and study problems. Rao et al. [3] have described how the IoT can function together and how cloud technology can solve big data problems. They have also demonstrated sensing as a service in the cloud domain with a few applications, such as increasing reality, agriculture, and environmental monitoring.

11.3   CBIT Architecture and Design Implementation

The CBIT has a different architectural layers, namely, application layer, a global cloud layer, and an IoT infrastructure layer. In the application layer, different application domains are used by the user through the different application protocols. In the cloud layer, IoT devices are interconnected with the cloud server. Data sensing and collection have been done by the IoT infrastructure's components.

11.3.1   Network Architecture of the CBIT

In Fig. 11.2, the network architecture of CBIT is presented. The assumption is that portable IoT users are usually grouped into personal organizations because they always move in the same trend at certain moments and places, such as the direction and speed [22]. IoT instruments enable interaction between us and the Internet. Therefore, once the transmission spectrum of two IoT customers is achieved, message bundles can be exchanged by them. In fact, IoT consumers at individual places in each cluster would produce duplicate/redundant packages, with a predominantly large possibility to report related activities. Lastly, IoT-restricted systems such as cameras, RFID tags, sensors, and intelligent meters would delegate highly complex calculations to the cloud to optimize performance.

11.3.2   Security and Privacy Authentication for the CBIT

For privacy-preserving authentication [9], we mainly focus on two aspects, the identity/location privacy protection and lightweight authentication solutions in the

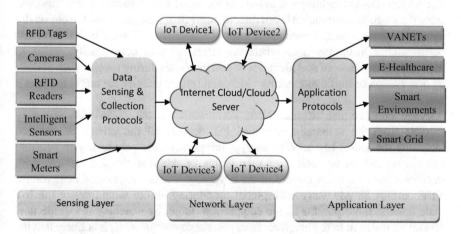

Fig. 11.2 Network architecture of CBIT [9]

CBIT. Conditional identity privacy is traditionally achieved by a group signature. However, the public key infrastructure (PKI) leads to the verification algorithm being inefficient and intolerable for the resource-constrained IoT devices due to the additional verification cost for the sender's public key certificate. To address the challenging security issue, an efficient privacy-preserving validation scheme save for location-based service (LBS) in the CBIT is proposed. Different from the existing work which saved the verification cost from the receiver's view, an efficient privacy-preserving LBS bundle filtering mechanism with dynamic social group formulation is novelty designed from the sender's aspect to simultaneously prevent duplicate LBS contents from aggregation.

11.3.3 Privacy-Preserving Testimonials in CBIT

We mainly focus on the security threats for CBIT, especially in the aspects of secure packet forwarding with outsourced aggregated transmission evidence generation and efficient privacy-preserving authentication with outsourced message filtering. Besides the traditional data confidentiality and enforceability, the unique security and privacy requirements in the CBIT are presented.

Identity privacy: Conditional identity privacy refers to the fact that the mobile IoT user's real identity should be well protected from the public; on the other hand, when some dispute occurs in emergency cases, it can also be effectively traced by the authority. The technique of pseudonyms has been widely adopted to achieve this target, but the periodically updated pseudonyms and certificates lead to intolerable computational cost for resource-constrained IoT nodes. More seriously, it cannot resist the physically dynamic tracing attack we identified for location privacy.

Location privacy: Location privacy seems especially critical in IoTs since the frequently exposed location privacy would disclose the living habit of the IoT user. The widely adopted technique is to hide its location through pseudonyms. However, since the location information is not directly protected, it cannot resist the physically dynamic tracing attack. Specifically, a set of malicious IoT users in collusion can be dispatched to the positions where the target IoT user occasionally visited, to physically record sets of real identities of passing nodes during specific time periods by observation or traffic monitoring video, and further identify the target IoT user's real identity.

Node compromise attack: Node compromise attack means the adversary extracts from the resource-constrained IoT devices all the private information including the secret key used to encrypt the packets, the private key to generate signatures, and so on, and then reprograms or replaces the IoT devices with malicious ones under the control of the adversary. The target-oriented compromise attack means an adversary with global monitoring ability would select the IoT node holding more packets as the compromise target by watching the traffic flow around all nodes in IoT. Therefore, from one single compromise, it is likely that the adversary obtains more packets for recovering the original message or impeding its successful delivery by the interruption.

Layer removing/adding attack: The layer removing attack occurs when a group of selfish IoT users removes all the forwarding layers between them to maximize their rewarded credits by reducing the number of intermediate transmitters sharing the reward. On the contrary, the layer adding attack means colluding IoT users maliciously detour the packet forwarding path between them for increased credits by increasing the total obtainable utility.

Forward and backward security: Due to the mobility and dynamic social group formulation in IoT, it is necessary to achieve forward and backward security. The former means that newly joined IoT users can only decipher the encrypted messages received after but not before they join the cluster, while the latter means that revoked IoT users can only decipher the encrypted messages before but not after leaving.

11.4 CBIT Applications

The CBIT introduced different types of smart applications to make things simple and rapid. It empowers us to simply connect, manage, and keep IoT information. CBIT platform also increases the scalability of our business model. Figure 11.3 represents various application areas embracing the CBIT approach. IoT cloud

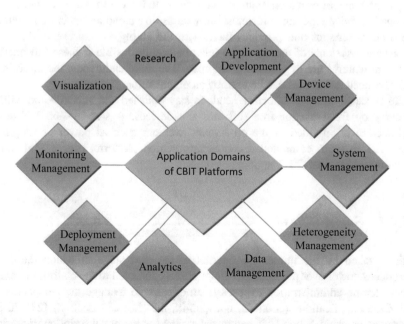

Fig. 11.3 Application domains of CBIT platforms [63]

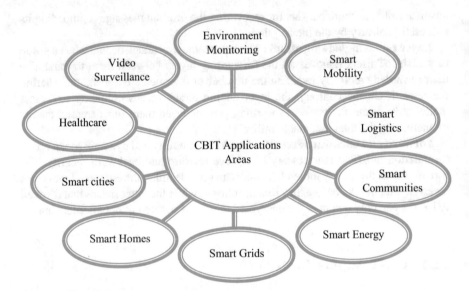

Fig. 11.4 CBIT applications in the different fields [43]

architecture relates to the various configurations that create up the cloud computing and information handling scheme of each corporation. A powerful cloud design helps facilitate information transfer with fresh IoT techniques. The Internet of Things is a daily experience for most individuals. IoT cloud architecture maintains these characteristics and provides them with the ability to improve further. The information systems of an enterprise must be highly flexible if these information handling features are to proceed to improve. The more endurance, the more fresh and enhanced techniques can be used by an organization.

To succeed, cloud architecture needs to stay polished and versatile and willing to carry on fresh friction-free software and devices. A step-by-step IoT cloud architecture is provided in the on-demand webinar such as power development using IoT in smart cities and real-time information platform. Figure 11.4 shows the different types of CBIT application areas.

11.4.1 Smart Cities

This is taken by the immersive Internet to the next stage in which the cloud introduces power over physical and digital items connected to the platform of smart cities. Traffic administration exposes Internet cams to manage the perspective of IoT Cloud applications (zoom-in, out, pan-in, the shift of direction) [23]. In the energy value chain, Smart Grid can reveal meters for power subscription triggering, disabling and closing control by retailers and customers. Smart cities are entitled by

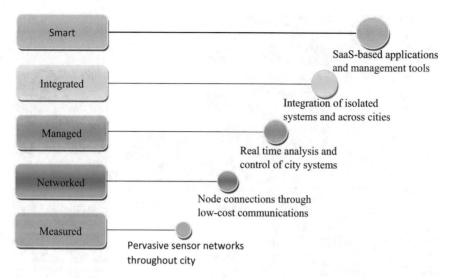

Fig. 11.5 Technology evaluation to a smart city [23]

the latest advances in significant technologies as low-cost communication, pervasive sensor networks, etc. Figure 11.5 shows technological evaluation in smart city areas applying the CBIT approach.

11.4.2 Smart Home

A large number of CBIT applications have automated home operations, enabling automation of internal activities (e.g., domestic safety control, smart metering, and power-saving) with the implementation of different portable devices and cloud computing. IoT, intelligent home, and cloud computing aren't just about a combination of technology. Instead, a balance should be established with the optimization of the consumption of resources between central and local computing. Computing can be performed or outsourced to the cloud on IoT and smart home devices. Where calculation is carried out will be determined by overhead trade, accessibility of information, reliance on the information, and the quantity and the safety aspects of transport information. On the one side, the cloud, IoT, and smart home three-way computation model should minimize the complete device costs, generally focused more on decreasing household energy consumption. On the other hand, the IoT and smart home-computing system should improve IoT clients to meet their requirements in cloud applications and tackle complex IoT, smart home, and cloud-based newly arisen problems.

Figure 11.6 represents and interconnects the primary sophisticated smart-home parts. On the left side, the smart home environment shows typical appliances linked

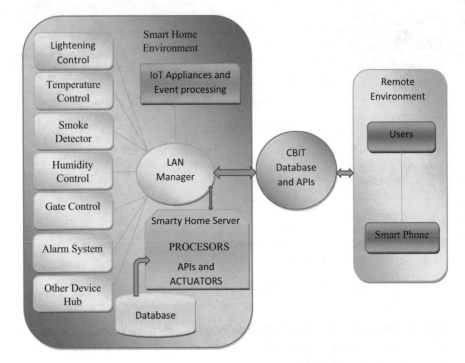

Fig. 11.6 CBIT-integrated smart homes [64]

to a local area network [LAN]. This allows interaction between the devices and outside of it. The smart home server and its database are connected to the LAN. The server monitors systems, logs activities, gives records, answer queries, and executes instructions. The smart home server sends information to the cloud for more extensive as well as prevalent assignments and remotely enables assignments using APIs. IoT home appliances also have Internet and LAN connections, which allow the smart home to comprise IoT. The Internet link enables the dwelling end-user to interact with the smart home to receive present data and to enable assignments remotely.

11.4.3 Smart Grids

CBIT can provide storage capacity to store IoT data for different types of smart grid applications. It can also support to manage and process data efficiently. Through the CBIT many interested researchers easily can get huge amounts of smart grid data for their research purpose. There are two types of CBIT smart grid applications, namely, existing application and potential application [24].

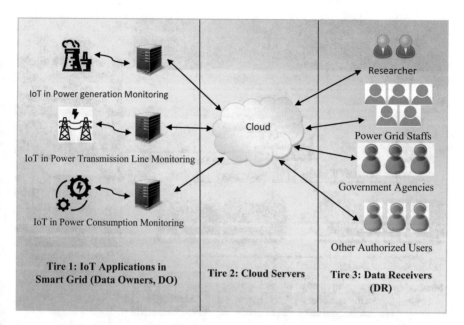

Fig. 11.7 Illustration of CBIT in smart grid [24]

CBIT is widely used in the smart grid. A huge quantity of information is collected and deposited in the back-end cloud servers by the IoT front-end devices. However, it is very important and difficult to achieve information integrity and device performance when acquiring and transmitting information. This challenging issue cannot be addressed well by current associated systems.

Figure 11.7 shows widespread information can be retrieved from various IoT workstations and analyzed by local front-end servers and then transmitted and saved on the cloud servers by evolving the CBIT. Different kinds of information consumers can access the information in the cloud. The grid employees can supervise the power grid performance continuously. The information for analysis or policymaking can be analyzed by researchers and government agencies.

11.4.4   Smart Energy

Energy is a very significant element for any community, industry, farming, and so forth. It is very essential for devices to manage energy effectively and thoughtfully. The energy usage is intermediately impacted by oil, coal, and so on energy production.

Growing economic growth and habits of consumption lead to increased energy demand. Because most of the energy production comes from renewable energy sources [25], this reduces the resource's cost of energy. Burning renewable energy

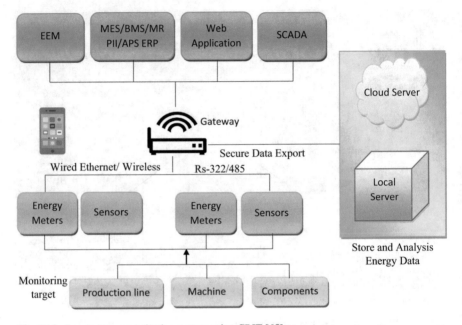

Fig. 11.8 Smart energy monitoring system using CBIT [65]

has improved carbon-di-oxide levels in the atmosphere, contributing to extreme weather conditions. Industries and businesses are therefore essential to adopt initiatives to decrease energy consumption, power efficiency, and cost reduction.

Cloud computing and IoT can perform together efficiently in order to ensure smart energy management, for example, intelligent meters, intelligent equipment, and renewable energy resources. As shown in Fig. 11.8, a particular energy monitoring system architecture can be obtained by generalizing best practices using the CBIT.

11.4.5   Smart Logistics

Logistics, transport, and warehousing are typically the first point of contact in integrated appliances that can be sensed and communicated decades before the word "IoT" has been invented. As per the worldwide logistics sector survey 2016, revealed by Technavio on 28 November [26], the associated global market in logistics is planned to increase at a compound annual growth rate (CAGR) of approx. 30% by 2020. Of course, the integrated logistics industry does not only concern the Industrial IoT in the toughest manner but also the entire environment that typically occurs in an embedded IoE strategy. It also includes facilities, privacy, the cloud, large amounts of data analysis, and various other third-party technology platforms. However, the IoT and the Internet of Industries, in general, play an

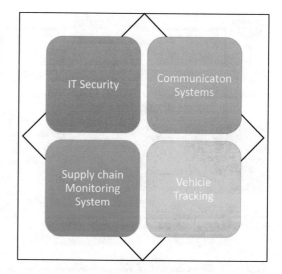

Fig. 11.9 Four pillars of connected logistics system [26]

Fig. 11.10 7 Rs of the smart logistics [26]

important part. Figure 11.9 shows that IT security, Communication platforms, vehicle tracking, and supply chain are the pillars of the interconnected smart logistics system. Figure 11.10 shows 7 Right (R) things in the logistics system.

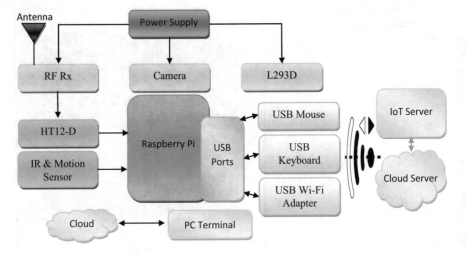

Fig. 11.11 Block diagram of video surveillance system [66]

11.4.6   Video Surveillance

Video surveillance [27] is the supervising process of an individual, a region, or a situation. This is usually done in an armed forces circumstance where border monitoring and adversary region are necessary for the safety of a nation. Human surveillance is carried out through the deployment of staff in nearly vulnerable fields to monitor alterations continuously. Figure 11.11 represents an overview of the video monitoring system's block diagram. However, people have their constraints and implementation is not always feasible in remote locations. The need for surveillance and monitoring has drastically increased over the course of the past few decades. Both private areas and public areas alike require these types of surveillance systems to ensure security and are generally very costly.

11.4.7   Environment Monitoring

Operations monitored [28] have recognized environment modifications and are used for a variety of reasons. Information collected during supervision should be conserved, evaluated, and related to the different pieces of data which alone can rank or influence the workplace in order to produce new and advanced smart systems. Figure 11.12 represents the environmental monitoring system adopting the CBIT approach.

The IoT has a vast as well as a continuous infrastructure of sensors concerning the environment and indicators in society today. This system monitors and controls important financial conditions, such as temperature, humidity, and CO amounts,

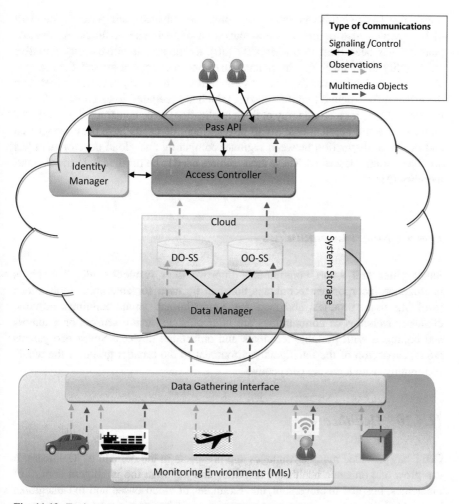

Fig. 11.12 Environment monitoring system in CBIT platform [28]

through sensors, and then transfers this data to the cloud. Everyone can use this data from any place on the web and display sensor data as graphical statistical data in a smartphone app.

11.4.8 Smart Mobility

The new paradigm and architecture of the IoT/IoE [29] allow the design and establishment of smart city infrastructures, but there are a number of problems to be solved on the contrary. As far as mobility is concerned, the towns that adopt the sensor age can use this disruptive technology to enhance their people's standard of lives which also points to a rationalization of their use of funds. A versatile platform has been built in Sii-Mobility [30], a domestic intelligent city initiative on mobility and transport. The manner the consumer travels around the city has altered dramatically over the last 20 years and the increasing interaction between the urban infrastructure and the sensors, which are prepared to catch up with their position, allows for a broad array of fresh applications to be constructed on fresh IoT architectures and applications. Smart development demands have been recognized and the logic distinction between regional computing and cloud development has also been briefly described. These architectures need to be fulfilled in terms of smart mobility [31].

11.4.9   Smart Communities

Smart cities and smart communities improve their resident's life. The city's intelligence incorporates techniques that can be used for enterprise applications involving smart products and facilities. Smart homes, smart buildings, airports, clinics universities or communities are interconnected using sensors or actuators and equipped with portable terminals and embedded devices. Smart ecosystems are an expansion of the intelligent environment to the broader group or the whole community from a private perspective.

11.4.10   Healthcare

CBIT has provided many advantages and prospects in the domain of healthcare. It can grow and enhance healthcare activities and maintain the intelligent control of medicines, hospital management, etc. Examples of cloud-based and IoT-integrated health facilities include the proper management of information, and the exchange of digital health care documents enables high-quality medical facilities and manage healthcare device data. Mobile phones are adapted for providing health data, safety, privacy, and reliability.

11.5   Analysis and Discussion

11.5.1   Challenges Faced in the CBIT Integration

Security and Privacy

CBIT enables information to be transported to the cloud from the actual world. One of the main problems still unsolved is how to contribute suitable authorization regulations and procedures while guaranteeing that only approved consumers have access to the confidential information; this is essential to the safeguarding of user privacy and, especially, to maintaining data integrity [32]. Furthermore, the absence of assurance in the service supplier, information about service level agreements (SLAs) [33], and physical data location [34] leads to problems when critical IoT devices or applications shift into the cloud.

Heterogeneity

The comprehensive heterogeneity of existing devices, platforms, operating systems, and facilities that could be used for fresh and advanced applications is a particular significant task confronted by the cloud-oriented IoT strategy. Cloud platforms are affected by problems of heterogeneity [35]. For example, cloud services usually have proprietary interface connectivity that allows the inclusion of resources depending on particular providers. In fact, when End Users embrace multi-cloud strategies, the heterogeneity problem may be intensified and thus services rely on several suppliers to enhance implementation efficiency and resilience.

Big Data

Big Data will achieve 50 billion IoT equipment by 2020; the enormous quantity of information that will be generated should be more carefully transported, accessed, saved, and processed. In light of recent advances in technology, it is clear that IoT will become a major source of Big Data. In addition to being subject to sophisticated analyses, the cloud can also enable cloud storage for a prolonged period of time [36]. The handling of the large quantity of information generated is an important problem, as when the entire output of the application is hugely dependent on the characteristics of this information management service. It is still a large problem to find an ideal alternative for information handling that will enable the cloud to process huge quantities of data. Moreover, data integrity is an essential component, not only for its impact on the quality of the service but also because of safety and security concerns [37].

Performance

A large number of information from IoT applications needs to be transferred to the cloud. As an outcome of this, the main problem is to obtain appropriate network performance to pass information to cloud platforms. This is because broadband development is still not consistent with the processing and computing development. Services and information available with elevated reactiveness should be accomplished in a variety of scenarios [7]. It is because thoroughness may be impaired by inconceivable things and real-time devices are highly susceptible to performances.

Legal Aspects

Legal aspects have been very important in the latest studies on certain applications. Service suppliers must adjust to different global provisions. In an attempt to add to the information set, clients should make donations. Many policies, assets, and organizations are accessible to support both the CSPs and their customers in addressing their legal issues [38]. However, since the regulations have evolved to adjust to the cloud, they will inevitably alter again along with further progress. This is as well the reality that cloud service providers often have to operate together to determine what wants to be accomplished by whom and when to comply with legal demands for customers that they serve. Both should acknowledge that measurements of data protection, probably the most important issue in most legislation and agreements, are not completely effective and that the possibility of failure is knowledgeable and minimized as fully as feasible.

Monitoring

Monitoring [39] is the key to efficiency, management of assets, capacity management, safety, SLAs, and debugging in cloud computing. The CBIT approach thus supports the same surveillance requirements from the cloud, but related problems are still affected by the speed, volume, and variety of IoT characteristics. Generic methods were developed to track computer devices without having to worry about the particular characteristics of each device sort. These approaches are commonly used for data on worldwide host resource retrieval schemes. However, generic solutions may not be appropriate for certain particular cloud characteristics, such as virtualization and strengthening of servers. In turn, cluster solutions and grid solutions have been designed for these special domains, but cloud-specific features also lack support. This is why cloud-specific alternatives need to be developed.

Large Scale

The IoT paradigm in the cloud allows new applications to be developed with the objective of integrating and analysis in IoT objects data from the real world. This calls for interaction with hundreds of millions of gadgets distributed in many areas [40]. The big extent of the subsequent schemes creates a number of fresh problems that are hard to resolve. For example, it is becoming hard to achieve computing capabilities and storage needs. In addition, the surveillance method has created it harder to distribute IoT phones as IoT systems experience connections and latency fluctuations.

11.5.2 Issues to Integrate the CBIT

Standardization

Many surveys have shown that the absence of expectations is crucial in connection with the CBIT's paradigm. This is important. While a percentage of the scientific society has introduced standardization [41] initiatives to implement IoT and cloud, it is decided that architectures, standard protocols, and APIs are needed to interconnect between heterogeneous intelligent things and to generate innovative services to comprise CBITs paradigm.

Fog Computing

Fog computing is a prototype that extends the cloud to the network's edge. Like the cloud, fog provides implementation facilities to customers [42]. In essence, fog could be regarded as an expansion of a cloud that works as crossroads between the edge of the network and the cloud. In fact, it acts with latency-sensitive apps, which allow other nodes to fulfill their delay requirements. Although memory, computation, and networking is both fog's and cloud's primary assets, the fog has some characteristics, such as geographically distributed place, and border locations and small latency. There are also massive servers, as opposed to the cloud that supports real-time communication and flexibility.

Cloud Capabilities

As with any interconnected environment, security is seen as a major problem in the CBIT. The IoT and the cloud face more possibilities of threats. In the IoT framework, encryption can ensure authentication, data protection, and reliability. However, anonymous source raids cannot be solved and the IoT on systems with restricted capacities [43] is also difficult to use.

SLA Enforcement

CBIT consumers have to provide data generated for transmission and processing, which is hard in some cases, based on application-dependent limitations. Making sure that cloud resources have a specific level of quality of service (QoS) due to a single supplier challenges some problems. Simultaneous cloud suppliers can be needed to prevent SLA [44, 45] violations. However, the most suitable mix of cloud suppliers is still an uncertain concern because of the heterogeneity of QoS management support, fees, and time.

Big Data

In the last chapter, we addressed Big Data [46, 47] as a major problem that is closely connected to the CBIT paradigm. Although several proposals have also been made, Big Data remains a critical, uncertain question and requires additional study. The CBIT requires the storage management and managing of enormous quantities of information from conservative areas and heterogeneous inputs; many features require complex functions to be done in real-time in CBIT systems.

Energy Efficiency

Current IoT applications in the cloud include periodic information transferred from IoT to the cloud that rapidly absorbs server resources. There are still important problems in the production of energy efficiency [48] in the handling and transmission of information. Various strategies to solve this problem have been suggested, like compression technology, effective data transmission, and information catching methods for reuse of gathered information.

Security and Privacy

While safety and privacy [49] are critical problems in studies that have gained much publicity, they continue to be transparent problems that need more attention. In fact, adaptation to various hacker threats remains a problem. In addition, another issue is offering suitable policies and practices on authorization, while maintaining that only approved consumers have access to delicate information.

11.5.3   Benefits of the CBIT

Communication

The application and data sharing of the CBIT frameworks are two important characteristics. Ubiquitous devices can be communicated via the IoT and technology can be used to make low-cost information allocation and compilation. The cloud is really an efficient and cost-effective option to communicate, implement, and monitor anything through embedded applications and custom websites. Fast systems are available to enable dynamic surveillance and remote control and real-time access to data. While the cloud may significantly evolve and promote IoT interconnections, in certain fields it also has limitations [50].

Storage

As IoT is used on billions of machines, the data source is large and generates huge semi-structured or not structured data [51]. It has three characteristics, namely big data: variety (e.g., information kinds), velocity (e.g., rate of information production), and volume (e.g., information size). In negotiating with the huge quantity of information generated by the IoT, the cloud is regarded to become one of the most price-effective and appropriate approaches. In addition, it offers fresh opportunities to integrate, aggregate, and share information with external parties.

Processing Capabilities

IoT systems have restricted computing capacities [11] that avoid on-site data processing and complicated information processing. The collected information is instead transmitted to highly functioning nodes; in fact, aggregation and handling are carried out in this respect. However, the achievement of scalability continues a task, without proper infrastructure. As a solution, the cloud offers an on-demand usage structure of virtual processing. Predictive algorithms and data-driven decision-making can be incorporated into the IoT to boost income and lower risk.

Scope

Hence, Billions of clients can connect through networks and collect a large amount of information the IoT can be expanded to the Internet of Everything (IoE) [52] environment, a network of networks with trillions of items that create fresh opportunities and hazards. The CBIT strategy offers innovative solutions and facilities oriented on the development of the cloud through IoT attributes, allowing the cloud to operate with such a series of real-world situations and leading to special features.

New Abilities

The heterogeneity of its devices, procedures, and techniques characterizes the IoT. So it may be very difficult to accomplish accuracy, scalability, interoperability, safety, accessibility, and effectiveness. Most of the problems are resolved by integrating IoT into the cloud [8].

New Models

IoT incorporation based on the cloud provides new instances for intelligent things, apps, and services. Such new prototypes are mentioned below:

- SaaS (Sensing as a Service) [53]: It enables access sensors information.

- EaaS (Ethernet as a Service) [43]: Its main role is to provide all-round remote device functionality.
- SAaaS (Sensing and Actuation as a Service) [54, 55]: that automatically supplies power logic.
- IPMaaS (Identity and Policy Management as a Service) [56]: It provides access to policies and the management of identities.
- DBaaS (Database as a Service) [57, 58]: It offers ubiquitous management of databases.
- SEaaS (Sensor Event as a Service) [59]: It sends notification facilities produced through the sensor activities.
- SenaaS (Sensor as a Service) [43]: It manages remote sensors;
- DaaS (Data as a Service) [60, 61]: It gives ubiquitous ingress to any data type.

11.5.4 Importance of the CBIT to the UAV

As the cloud has a massive computing capacity, most of the UAV or drone information analyses on the server can be done instead of the UAVs, which decreases energy consumption. In contrast to the minimal UAV space, CBIT offers comprehensive and flexible computing facilities. It improves information consistency with data backup, allowing the ability to access past log records even if the UAVs are out of operation. The data recovery is therefore enhanced.

In addition, UAV flight control methods [62] should be used for real-time implementation, path-planning management, and collision avoidance. Reliable Internet service is another consideration of the UAV. UAV allows constant network synchronization to reach the Internet and its services across its APIs. The premise of a trustworthy correlation is valid for services in city areas like smart cities; otherwise, the connection infrastructure for the operation of UAV should be provided to the operating area. Therefore, UAV facilities are real-world systems that enable the natural atmosphere to be sensed and impacted.

Drone [67] plays crucial role in the different types of area. Drone has potential to provide quick service to the smart cities autonomously. It can give faster serves in smart cities like goods delivery, traffic monitoring, land-site visiting, fire servicing, and pollution control. It can collect information from cloud and manage by the IoT device. IoT device serves various facilities like drone tracking, minimal path setting, and drone security management. It will be very well-being if CBIT connect with drone. It will give secure, scalable, and reliable services.

Drone can be used to increase privacy and protection in individual or group of smart home. It can play role for cloud services through IoT technology without a smart phone or smart devices. It also serves newspaper, magazine, or something like that which are most needed in daily life door to door at any time. It will be very useful if drone can act combining with CBIT applications.

Drone can be used in smart grid utilities like power line inspection, wind turbine inspection, magnetic area detection, thermal image capturing, and firefighting tools.

It can maintain or check electricity transmission line quickly. It can take any images from smart grid environment at any time. So, it will be very useful in these area within a very less time.

Drone provides different type of applications when it is needed. It can check transmission power spot check, routine maintenance, substation upgrades, and storm restoration in the transmission and distribution application. It also provides other services like site planning, construction, and asset transfer in the wind and solar energy application.

One of the useful applications of drone combining CBIT is supply chain management. It can also be used for environment monitoring like plant control, pests control, field observation, flood assessment, and river mapping.

11.6   Conclusion and Future Scope

The technology of cloud computing provides many opportunities, but it also has various constraints. Cloud computing relates to an environment that stores information and processes information outside the portable system. Cloud computing's input to the IoT technology demonstrates how cloud computing innovation increases the IoT operation. Cloud development trends seem to favor computer leadership, information leadership, and application-based models. There is also evidence that a certain IoT cloud can serve multiple applications. There are very fewer research fields to develop in order to provide a better understanding of the scientific community of conducting real-life IoT-based tests. The main areas covered by present IoT windows also are application advancement and business tracking. However, the IoT applications do not only have to provide strict control in aspects of context awareness, large information processing, and device leadership problems but are also close to the specified varieties of ambiguity.

The cloud systems are ingredients that provide IoT with present and essential demands, such as real-time processing, flexible saving, and worldwide access, and expand to other possibilities such as computer training. The cloud systems have distinct goals to help maintain, track, and deploy functions in distinct classifications such as batch processing, distributed databases, distributed queues, and real-time processing. The cloud architecture offers IoT and cloud systems with the necessary assets for saving, networking, and computation. Finally, the IoT software offers the fundamental IoT systems an object model and the framework for communicating with cloud. The privacy and security issues remain in the IoT and cloud computing and there is an absence of standardization in these two fields. In future studies, the efficacy of the CBIT strategy will be tested in multiple applications such as UAV's application and healthcare management.

References

1. J.D. Bokefode, A.S. Bhise, P.A. Satarkar, D.G. Modani, Developing a secure cloud storage system for storing IoT data by applying role-based encryption. Proc. Comp. Sci. **89**, 43–50 (2016). https://doi.org/10.1016/j.procs.2016.06.007
2. Y. Liu, B. Dong, B. Guo, J. Yang, W. Peng, Combination of cloud computing and internet of things (IoT) in medical monitoring systems. Int. J. Hybrid Inf. Technol. **8**, 367–376 (2015). https://doi.org/10.14257/ijhit.2015.8.12.28
3. B.B.P. Rao, P. Saluia, N. Sharma, A. Mittal, S.V. Sharma. *Cloud Computing for Internet of Things & Sensing-Based Applications*, 2012 Sixth International Conference on Sensing Technology (ICST), Kolkata (2012), pp 374–380. doi: https://doi.org/10.1109/ICSensT.2012.6461705
4. I. Odun-Ayo, M. Ananya, F. Agono, R. Goddy-Worlu. *Cloud Computing Architecture: A Critical Analysis* (Melbourne, 2018), pp 1–7. DOI: https://doi.org/10.1109/ICCSA.2018.8439638
5. M. Aazam, I. Khan, A.A. Alsaffar, E.N. Huh *Cloud of Things: Integrating Internet of Things and Cloud Computing and the Issues Involved*, Proceedings of 2014 11th International Bhurban Conference on Applied Sciences & Technology (IBCAST) (Islamabad, 2014), pp 414–419
6. J. Barbaran, M. Diaz, B. Rubio. *A virtual channel-based framework for the integration of wireless sensor networks in the cloud*. In: Proceedings of the 2nd International Conference on Future Internet of Things and Cloud (FiCloud-2014) (Barcelona, 2014), pp 334–339. DOI: https://doi.org/10.1109/FiCloud.2014.59
7. C. Stergiou, K.E. Psannis, B.G. Kim, B.B. Gupta, Secure integration of IoT and cloud computing. Fut. Gener. Comput. Syst. **78**, 964–975 (2018). https://doi.org/10.1016/j.future.2016.11.031
8. P.P. Ray, A survey of IoT cloud platforms. Fut. Comp. Inform. J. **1**(1–2), 35–46 (2016). https://doi.org/10.1016/j.fcij.2017.02.001
9. J. Zhou, Z. Cao, X. Dong, A.V. Vasilakos, Security and privacy for CBIT: Challenges. IEEE Commun. Mag. **55**(1), 26–33 (2017). https://doi.org/10.1109/MCOM.2017.1600363CM
10. M. Díaz, C. Martín, B. Rubio, State-of-the-art, challenges, and open issues in the integration of internet of things and cloud computing. J. Netw. Comput. Appl. **67**, 99–117 (2016). https://doi.org/10.1016/j.jnca.2016.01.010
11. S. Sivakumar, V. Anuratha, S. Gunasekaran, Survey on integration of cloud computing and internet of things using application perspective. Int. J. Emerg. Res. Manag. Technol. **6**, 101–108 (2017). https://doi.org/10.23956/ijermt/SV6N4/101
12. M.R. Belgaum, S. Soomro, Z. Alansari, S. Musa, M. Alam, M.M. Suud. *Challenges: Bridge Between Cloud and IoT*, In 2017 4th IEEE International Conference on Engineering Technologies and Applied Sciences (ICETAS), (Salmabad, 2017), pp. 1–5. doi: https://doi.org/10.1109/ICETAS.2017.8277844
13. A.Auger, E. Exposito, E. Lochin. *Sensor Observation Streams Within CBIT Platforms: Challenges and Directions*, 2017 20th Conference on Innovations in Clouds, Internet and Networks (ICIN) (Paris, 2017), pp. 177–184. doi: https://doi.org/10.1109/ICIN.2017.7899407
14. S.M. Babu, A.J. Lakshmi, B.T. Rao. *A Study on Cloud Based Internet of Things: CloudIoT*, 2015 Global Conference on Communication Technologies (GCCT) (Thuckalay, 2015), pp 60–65. doi: https://doi.org/10.1109/GCCT.2015.7342624
15. A. Botta, W. de Donato, V. Persico, A. Pescapé, Integration of cloud computing and internet of things: A survey. Futur. Gener. Comput. Syst. **56**, 684–700 (2016). https://doi.org/10.1016/j.future.2015.09.021
16. K.D. Chang, C.Y. Chen, J.L. Chen, H.C. Chao. *Internet of Things and Cloud Computing for Future Internet, Security-Enriched Urban Computing and Smart Grid* (2011), pp. 1–10. doi:https://doi.org/10.1007/978-3-642-23948-9_1
17. T. Choudhury, A. Gupta, S. Pradhan, P. Kumar, Y.S. Rathore. Privacy and security of cloud-based internet of things (IoT), 2017 3rd International Conference on Computational Intelligence and Networks (CINE) (Odisha, 2017), pp. 40–45. doi: https://doi.org/10.1109/

CINE.2017.28

18. A.F. Da Silva, R.L. Ohta, M.N. dos Santos, A.P.D. Binotto, A cloud-based architecture for the internet of things targeting industrial devices remote monitoring and control. IFAC-Papers OnLine **49**(30), 108–113 (2016). https://doi.org/10.1016/j.ifacol.2016.11.137

19. H. Elazhary, Internet of things (IoT), mobile cloud, cloudlet, mobile IoT, IoT cloud, fog, mobile edge, and edge emerging computing paradigms: Disambiguation and research directions. J. Netw. Comput. Appl. **128**, 105–128 (2018). https://doi.org/10.1016/j.jnca.2018.10.021

20. I.F. Akyildiz, J.M. Jornet, The internet of nano-things. IEEE Wirel. Commun. **17**(6), 58–63 (2010). https://doi.org/10.1109/MWC.2010.5675779

21. M.C. Domingo, An overview of the Internet of underwater things. J. Netw. Comput. Appl. **35**(6), 1879–1890 (2012). https://doi.org/10.1016/j.jnca.2012.07.012

22. J. Zhou, X. Dong, Z. Cao, A.V. Vasilakos, Secure and privacy preserving protocol for cloud-based vehicular DTNs. IEEE Trans. Info. Forens. Secur. **10**(6), 1299–1314 (2015). https://doi.org/10.1109/TIFS.2015.2407326

23. B. Kommadi. *Smart Cities: IoT Cloud* (2016). https://iasaglobal.org/smart-cities-iot-cloud/. Accessed 20 Aug 2019

24. Z. Guan, J. Li, L. Wu, Y. Zhang, J. Wu, X. Du, Achieving efficient and secure data acquisition for cloud-supported Internet of things in smart grid. IEEE Internet Things J. **4**(6), 1934–1944 (2017). https://doi.org/10.1109/jiot.2017.2690522

25. V. Mani, Abhilasha, Gunasekhar, Lavanya, Iot based smart energy management system. Int. J. Appl. Eng. Res. **12**(16), 5455–5462 (2017)

26. I-scoop. Connected logistics 2017–2020: IoT, cloud and analytics as key drivers (2017), https://www.i-scoop.eu/digital-transformation/transportation-logistics-supply-chain-management/connected-logistics-2017-2020/. Accessed 15 Aug 2019

27. R.M. Patil, R. Srinivas, Y. Rohith, N.R. Vinay, D. Pratiba. *IoT Enabled Video Surveillance System Using Raspberry Pi*, 2017 2nd International Conference on Computational Systems and Information Technology for Sustainable Solution (CSITSS) (Bangalore, 2017), pp 1–7. https://doi.org/10.1109/CSITSS.2017.8447877

28. M. Fazio, A. Celesti, A. Puliafito, M. Villari, Big data storage in the cloud for smart environment monitoring. Proc. Comp. Sci. **52**, 500–506 (2015). https://doi.org/10.1016/j.procs.2015.05.023

29. K.S. Yeo, M.C. Chian, T.C. Wee Ng, D.A. Tuan. *Internet of Things: Trends, Challenges and Applications*, In 2014 International Symposium on Integrated Circuits (ISIC) (Singapore, 2014), pp 568–571. doi: https://doi.org/10.1109/ISICIR.2014.7029523

30. C. Badii, P. Bellini, A. Difino, P. Nesi, Sii-mobility: An IoT/IoE architecture to enhance Smart City mobility and transportation services. Sensors **19**(1), 1 (2018). https://doi.org/10.3390/s19010001

31. M. Barcelo, A. Correa, J. Llorca, A.M. Tulino, J.L. Vicario, A. Morell, IoT-cloud service optimization in next generation smart environments. IEEE J. Select. Areas Commun. **34**(12), 4077–4090 (2016). https://doi.org/10.1109/JSAC.2016.2621398

32. Y.W. Lee, L. Pipino, D.M. Strong, R.Y. Wang, Process-embedded data integrity. J. Database Manag. **15**(1), 87–103 (2004). https://doi.org/10.4018/jdm.2004010104

33. J.J.M. Trienekens, J.J. Bouman, M.V.D. Zwan, Specification of service level agreements: Problems, principles and practices. Softw. Qual. J. **12**(1), 43–57 (2004). https://doi.org/10.1023/b:sqjo.0000013358.61395.96

34. M. Irain, J. Jorda, Z. Mammeri, Landmark-based data location verification in the cloud: Review of approaches and challenges. J. Cloud Comp. **6**(31), 20 (2017). https://doi.org/10.1186/s13677-017-0095-y

35. R. Boutaba, L. Cheng, Q. Zhang, On cloud computational models and the heterogeneity challenge. J. Int. Ser. Appl. **3**(1), 77–86 (2011). https://doi.org/10.1007/s13174-011-0054-7

36. A. Botta, W.D. Donato, V. Persico, A. Pescapé. On the Integration of Cloud Computing and Internet of Things, 2014 International Conference on Future Internet of Things and Cloud (Barcelona, 2014), pp. 23–30. doi: https://doi.org/10.1109/FiCloud.2014.14

37. N. Mathur, R. Purohit, Issues and challenges in convergence of big data, cloud and data science. Int. J. Comp. Appl. **160**(9), 7–12 (2017). https://doi.org/10.5120/ijca2017913082

38. D.G. Gordon *Legal Aspects of Cloud Computing, Encyclopedia of Cloud Computing* (2016), pp 462–475. doi:https://doi.org/10.1002/9781118821930.ch38
39. G.D.C. Rodrigues, R.N. Calheiros, V.T. Guimaraes, G.L.D. Santos, M.B. de Carvalho, L.Z. Granville, R. Buyya. *Monitoring of Cloud Computing Environments*, In Proceedings of the 31st Annual ACM Symposium on Applied Computing—SAC '16 (2016), pp. 378–383. doi:https://doi.org/10.1145/2851613.2851619
40. L. Luo, S. Meng, X. Qiu, Y. Dai, Improving failure tolerance in large-scale cloud computing systems. IEEE Trans. Reliab. **68**(2), 620–632 (2019). https://doi.org/10.1109/TR.2019.2901194
41. J. Saleem, M. Hammoudeh, U. Raza, B. Adebisi, R. Ande. *IoT standardization*, In Proceedings of the 2nd International Conference on Future Networks and Distributed Systems—ICFNDS '18, USA (2018), vol. 1, pp. 9. doi:https://doi.org/10.1145/3231053.3231103
42. M. Gomes, M.L. Pardal, Cloud vs fog: Assessment of alternative deployments for a latency-sensitive IoT application. Proc. Comp. Sci. **130**, 488–495 (2018). https://doi.org/10.1016/j.procs.2018.04.059
43. H.F. Atlam, A. Alenezi, A. Alharthi, R.J. Walters, G.B. Wills. *Integration of Cloud Computing with Internet of Things: Challenges and Open Issues*, In 2017 IEEE International Conference on Internet of Things (iThings) and IEEE Green Computing and Communications (GreenCom) and IEEE Cyber, Physical and Social Computing (CPSCom) and IEEE Smart Data (SmartData), (Exeter, 2017), pp 670–675. doi: https://doi.org/10.1109/iThings-GreenCom-CPSCom-SmartData.2017.105
44. S. Mubeen, S.A. Asadollah, A.V. Papadopoulos, M. Ashjaei, H. Pei-Breivold, M. Behnam, Management of Service Level Agreements for cloud services in IoT: A systematic mapping study. IEEE Access **6**, 30184–30207 (2018). https://doi.org/10.1109/access.2017.2744677
45. A. Alzubaidi, E. Solaiman, P. Patel, K. Mitra, Blockchain-based SLA Management in the Context of IoT. IT Professional **21**(4), 33–40 (2019). https://doi.org/10.1109/MITP.2019.2909216
46. N.K. Suchetha, H.S. Guruprasad, Integration of IOT, cloud and big data. Glob. J. Eng. Sci. Res. **2**(7), 251–258 (2015)
47. H. Cai, B. Xu, L. Jiang, A.V. Vasilakos, IoT-based big data storage systems in cloud computing: Perspectives and challenges. IEEE Internet Things J. **4**, 75–87 (2016). https://doi.org/10.1109/JIOT.2016.2619369
48. H.M. Al-Kadhim, H.S. Al-Raweshidy, Energy efficient and reliable transport of data in CBIT. IEEE Access **7**, 64641–64650 (2019). https://doi.org/10.1109/ACCESS.2019.2917387
49. Z. Qureshi, N. Agrawal, D. Chouhan, Cloud based IOT: Architecture, application, challenges and future. Int. J. Sci. Res. Comp. Sci. Eng. Inform. Technol. **3**(7), 359–368 (2018)
50. B.R. Ferrer, W.M. Mohammed, E. Chen, J.L.M. Lastra. *Connecting Web-Based IoT Devices to a Cloud-Based Manufacturing Platform*, In IECON 2017—43rd Annual Conference of the IEEE Industrial Electronics Society (Beijing, 2017), pp. 8628–8633. doi: https://doi.org/10.1109/IECON.2017.8217516
51. H. Khan, NoSQL: A database for cloud computing. Int. J. Comp. Sci. Networ. **3**(6), 498–501 (2014)
52. M.H. Miraz, M. Ali, P.S. Excell, R. Picking. *A Review on Internet of Things (IoT), Internet of Everything (IoE) and Internet of Nano Things (IoNT)* (North East Wales, 2015), pp. 219–224. doi:https://doi.org/10.1109/ITechA.2015.7317398
53. Y.R. S Kumar, H.N. Champa, An extensive review on sensing as a service paradigm in IoT: Architecture, research challenges, lessons learned and future directions. Int. J. Appl. Eng. Res. **14**(6), 1220–1243 (2019)
54. S. Satpathy, B. Sahoo, A.K. Turuk, Sensing and actuation as a service delivery model in cloud edge centric internet of things. Futur. Gener. Comput. Syst. **86**, 281–296 (2018). https://doi.org/10.1016/j.future.2018.04.015
55. S. Satpathy, B. Sahoo, A.K. Turuk, P. Maiti. *Sensing–Actuation as a Service in Cloud Centric Internet of Things, Lecture Notes in Networks and Systems* (2017), pp. 57–66. doi:https://doi.org/10.1007/978-981-10-5523-2_6

56. M. Zhou, R. Zhang, D. Zeng, W. Qian. *Services in the Cloud Computing Era: A Survey*, In 2010 4th International Universal Communication Symposium, Beijing (2010), pp. 40–46. doi: https://doi.org/10.1109/IUCS.2010.5666772
57. W.A.L. Shehri, Cloud database database as a service. Int. J. Data. Manag. Syst. **5**, 1–12 (2013). https://doi.org/10.5121/ijdms.2013.5201
58. S.D. Bijwe, P.L. Ramteke, Database in cloud computing database-as-a service (DBaas) with its challenges. Int. J. Comput. Sci. Mob. Comput. **4**(2), 73–79 (2015)
59. S.K. Dash, S. Mohapatra, P.K. Pattnaik, A survey on applications of wireless sensor network using cloud computing. Int. J. Comp. Sci. Emerg. **1**(4), 50–55 (2010)
60. O. Terzo, P. Ruiu, E. Bucci, F. Xhafa. *Data as a Service (DaaS) for Sharing and Processing of Large Data Collections in the Cloud*, 2013 Seventh International Conference on Complex, Intelligent, and Software Intensive Systems (Taichung, 2013), pp. 475–480. doi: https://doi.org/10.1109/CISIS.2013.87
61. X. Wang, L.T. Yang, H. Liu, M.J. Deen, A big data-as-a-service framework: State-of-the-art and perspectives. IEEE Trans. Big Data **4**(3), 325–340 (2018). https://doi.org/10.1109/TBDATA.2017.2757942
62. S. Mahmoud, N. Mohamed, J. Al-Jaroodi, Integrating UAVs into the cloud using the concept of the web of things. J. Robot., 1–10 (2015). https://doi.org/10.1155/2015/631420
63. P.P. Ray, A survey on internet of things architectures. J. King Saud Univer. Comp. Inform. Sci. **30**(3), 291–319 (2018). https://doi.org/10.1016/j.jksuci.2016.10.003
64. M. Domb. *Smart Home Systems Based on Internet of Things, Israel* (2019). https://doi.org/10.5772/intechopen.84894
65. F. Shrouf, G. Miragliotta, Energy management based on internet of things: Practices and framework for adoption in production management. J. Clean. Prod. **100**, 235–246 (2015). https://doi.org/10.1016/j.jclepro.2015.03.055
66. H. Pandhi. *IoT Projects: Wireless Video Surveillance Robot using Raspberry Pi* (2017), https://electronicsforu.com/electronics-projects/iot-video-surveillance-raspberry-pi. Accessed 15 Aug 2019
67. A. Koubaa, B. Qureshi, DroneTrack: Cloud-based real-time object tracking using unmanned aerial vehicles. IEEE Access **6**, 13810–13824 (2018). https://doi.org/10.1109/access.2018.2811762

Debashis Das is currently studying as a university research scholar at Kalyani University. He has completed the Master of Technology in Computer Science and Engineering from Kalyani Government Engineering College. He has completed Bachelor of Technology in Computer Science and Engineering from Government College of Engineering and Leather Technology. His research interests are including Cloud Computing, IoT, and Blockchain.

Sourav Banerjee holds a Ph.D. degree from the University of Kalyani in 2017. He is currently an Assistant Professor at Department of Computer Science and Engineering of Kalyani Government Engineering College at Kalyani, West Bengal, India. He has authored numerous reputed journal articles, book chapters, and international conferences. His research interests include Big Data, Cloud Computing, Cloud Robotics, Distributed Computing, and Mobile Communications, IoT. He is a member of IEEE, ACM, IAE, and MIR Labs as well. He is a SIG member of MIR Lab, USA. He is an Editorial board member of Wireless Communication Technology.

Utpal Biswas received his B.E., M.E., and Ph.D. degrees in Computer Science and Engineering from Jadavpur University, India, in 1993, 2001, and 2008, respectively. He served as a faculty member in NIT, Durgapur, India, in the Department of Computer Science and Engineering from 1994 to 2001. Currently, he is working as a professor in the Department of Computer Science and Engineering, University of Kalyani, West Bengal, India. He has over 130 research articles in different journals, book chapters, and conferences. His research interests include optical communication, ad hoc and mobile communication, semantic web services, E-governance, Cloud Computing, etc.

Index

Printed in the United States
by Baker & Taylor Publisher Services